RECENT ADVANCES IN PRIMATOLOGY

Volume Two
CONSERVATION

RECENT ADVANCES IN PRIMATOLOGY

Volume 1   Behaviour
Volume 2   Conservation
Volume 3   Evolution
Volume 4   Medicine

RECENT ADVANCES IN PRIMATOLOGY

Volume Two
CONSERVATION

Edited by

D. J. CHIVERS
*Sub-Department of
Veterinary Anatomy,
University of Cambridge,
Cambridge, U.K.*

W. LANE-PETTER
*Clinical School Animal House,
New Addenbrookes Hospital,
Hills Road,
Cambridge, U.K.*

1978

ACADEMIC PRESS
London . New York . San Francisco

A Subsidiary of Harcourt Brace Jovanovich, Publishers

ACADEMIC PRESS INC. (LONDON) LTD.
24/28 Oval Road,
London NW1

United States Edition published by
ACADEMIC PRESS INC.
111 Fifth Avenue
New York, New York 10003

Copyright © 1978 by
ACADEMIC PRESS INC. (LONDON) LTD.

All Rights Reserved
No part of this book may be reproduced in any form by photostat, microfilm, or any other
means, without written permission from the publishers

Library of Congress Catalog Card Number: 77-75366
ISBN: 0-12-173302-5

Printed in Great Britain by
Galliard (Printers) Ltd, Great Yarmouth

LIST OF CONTRIBUTORS

Altmann, D., 5001 Erfurt/GDR, Thuringer Zoopark, Zurin Zoopark 8-10.
Angst, W., Zoological Institute, Rheinsprung 9, CH-4051 Basel, Switzerland.
Asibey, E.O.A., Department of Game & Wildlife, PO Box L239, Accra, Ghana.
Awunti, Hon. Joseph, Ministry of Agriculture, Yaoundé, Cameroun.
Baker, B.A., MRC Dunn Nutritional Laboratory, Milton Road, Cambridge, UK
Birkner, F.E., New York University Medical Center, 400 E. 34th Street, New York, NY 10016, USA.
Bleby, J., MRC Laboratory Animals Centre, Woodmansterne Road, Carshalton, Surrey SM5 4EF, UK
Borner, Monica, Seestrasse 167, CH-8800 Thalwil, Switzerland.
Bourne, G., Yerkes Primate Center, Emory University, Atlanta, Ga. 30322, USA.
Brotoisworo, Edy, Biology Faculty, University of Gadjah Mada, Sekip Utara, Yogyakarta, Indonesia.
Burton, J.A., Fauna Preservation Society, c/o London Zoological Society, Regent's Park, London NW1 4RY, UK
Chernyshev, V.I., USSR AMS Institute of Poliomyelitis and Viral Encephalitis, Moscow, USSR.
Chivers, D.J., Sub-Department of Veterinary Anatomy, Tennis Court Road, Cambridge CB2 1QS.
Colillas, O.J., Uruguay 181, Punta Chica, Victoria, Pcia, Buenos Aires, Argentina.
Cummins, L.B., Southwest Foundation for Research and Education, PO Box 28147, San Antonio, Texas 78284, USA.
Darsono, C.L., PO Box 1152, Jakarta-Barat, Indonesia.
Dukelow, W.R., Endocrine Research Unit, Michigan State University, East Lansing, Michigan 48842, USA.
Eudey, Ardith, A., Department of Anthropology, University of California, Davis, California 95616, USA.
Fitter, R.S.R., Fauna Preservation Society, c/o Zoological Society of London, Regent's Park, London NW1 4RY, UK
Fitzgerald, T.A., New York University Medical Center, 550 First Avenue, New York, NY 10016, USA.
Goodall, A.G., 7 Stirling Drive, Bearsden, Glasgow G61 4NX, Scotland, UK
Hiddleston, W.A., ICI Pharmaceuticals Division, Mereside, Alderley Edge, Macclesfield, Cheshire, UK.
Joss, G.E., Waysmeet, 8 White Post Mill, Redhill, Surrey, UK.
Korte, R., PZM Primatenzentrum Münster KG, Vorbergweg 41, D-4400 Münster, Germany.
Lane-Petter, W., (Retired), Clinical School Animal House, Addenbrooke's

List of Contributors

Hospital, Hills Road, Cambridge, UK.

Laursen, A.C., 583 Jalan 17/17, Petaling Jaya, Malaysia.

Lunn, S.F., MRC Reproductive Biology Unit, 2 Forrest Road, Edinburgh EH1 2QW, Scotland.

Mahoney, C.J., Wisconsin Regional Primate Research Center, Madison, Wisconsin 53706, USA.

Marsh, Clive, Department of Psychology, Bristol University, 8-10 Berkeley Square, Bristol BS8 1HH, UK

Mohamed Khan bin Momin Khan, IBU Pejabat Jabatan Mergasetua, PO Box 611, Kuala Lumpur, Malaysia.

Mohnot, S.M., Department of Zoology, University of Jodhpur, Jodhpur, India.

Moor-Jankowski, J., New York University Medical Center, 550 First Avenue, New York, NY 10016, USA.

Muckenhirn, Nancy A., LIAR-NRC, 2101 Constitution Avenue, Washington DC 20418, USA.

Pucak, G.J., Charles River Breeding Labs, Inc., 251 Ballardvale Street, Wilmington, Massachusetts 01887, USA.

Schwaier, Anita, Battelle Institute, Department Experimental Medicine, 6 Frankfurt/Main, Am Römerhof 35, BRD.

Socha, W.W., New York University Medical Center, 550 First Avenue, New York, NY 10016, USA.

Struhsaker, T.T., PO Box 409, Fort Portal, Uganda, East Africa.

Thorington, R.W., National Museum of Natural History, Smithsonian Institution, Washington, DC 20560, USA.

Tribe, G.W., Shamrock Farms (G.B.) Ltd., Upper Horton Farm, Small Dole, Henfield, Sussex, UK.

Trollope, J., Department of Growth & Development, Institute of Child Health, 30 Guildford Street, London WC1, UK

Turton, J.A., MRC Laboratory Animals Centre, Woodmansterne Road, Carshalton, Surrey SM5 4EF, UK

Valerio, D.A., Hazleton Laboratories America Inc., 9200 Leesburg Turnpike, Vienna, Virginia 22180, USA.

Vanzolini, P., Museum de Zoologia, Universidade de Sao Paulo, Caixa Postal 7172, Sao Paulo, Brazil.

van Vreeswijk, W., Primate Center TNO, 151 Lange Kleiweg, Rijswijk ZH, Netherlands.

PREFACE

The Congresses of the International Primatological Society, held biennially since 1966, are noted for their social success- for the enjoyment and mutual benefits of mixing scientists from diverse disciplines, who share an interest in the one order of animals that is rather special for man. It is these very differences in training and specific interests in primates, however, which hinder the real academic success of these Congresses.

When starting to plan the Sixth Congress three years ago, we wondered whether to continue the tradition of a programme of very varied topics, with the Congress inevitably fragmenting into specialist groups, or to attempt to focus the multi-disciplinary interests on a particular theme, since primatology had expanded so much. The latter might still be the more constructive challenge for the future, but we opted for the traditional format (with modifications), because we felt unable to attract the whole range of primatologists for any of the themes proposed.

Nevertheless, we resolved, in the sessions we would convene on specific topics in primatology, to ensure (1) a real, broad synthesis of recent advances, with a careful selection of papers, (2) the inclusion of review papers of general interest near the start of each session, and (3) the exploration of each topic with increasing informality (and specialization) - from symposia, to short-paper sessions, to round-table discussions - with ample time for discussion.

While soliciting topics and papers, care was taken to avoid an excessive overlap of sessions. The resulting five-day programme was derived almost entirely from the interests expressed by prospective delegates, with some bias towards local interests.

Nearly 600 delegates attended the Sixth Congress of the International Primatological Society between August 22nd and 28th 1976 on the Sidgwick Avenue Site of the University of Cambridge, England. Nearly half the delegates presented papers, and an extensive selection of these is published here. In editing the four volumes which comprise the proceedings of the Congress it has been more necessary than in compiling the programme to select really relevant material. The aim has been to produce thorough syntheses of recent developments in the fourteen or so areas discussed, rather than just another collection of congress papers.

Thus, the bulk of the four volumes is composed of the fourteen symposia - 6 on Behaviour, 3 on Conservation, 2 on Evolution and 3 on Medicine - with important supplementary material from 7 short-paper sessions (in the Behaviour and Evolution volumes) and from 6 (out of the 8 convened) round-table discussions.

Preface

Volume 1 contains papers relating mainly to Behaviour - inter-individual relations and group structure, early social behaviour (mother-infant relations and play), demography and social organization, behavioural terminology, feeding behaviour in relation to food availability and composition, sexual and aggressive behaviour (including sexuality in apes), motor coordination, hearing and acoustic communication, language and its origins, cognition and learning, and aspects of the visual system.

Volume 2 contains papers relating mainly to Conservation - current problems in primate conservation in Africa, Asia and the Americas, the trade and supply of primates, and breeding primates in captivity.

Volume 3 contains papers relating mainly to Evolution - on primate evolution in general, and molecular and chromosomal evolution in particular, on behavioural factors in prosimian evolution, the phylogeny of tarsiers, S.E. Asian primates, methods of phylogenetic inference, and hominid evolution.

Volume 4 contains papers relating mainly to Medicine - infectious diseases including zoonoses, research on transmissible cancer, and the use of primates in research on human reproduction including fertility control.

Special thanks are due to those who contributed to the financial success of the Congress:- All-type Tools (Woolwich), Boots Company Ltd., Barclays International, British Council, Cambridge City, Cambridge University, Commonwealth Foundation, Hope Farms of Holland, Huntingdon Research Centre, I.C.I., I.U.B.S., Labusure Animal Foods, Shamrock Farms (GB) Ltd., Roche Products Lts., The Royal Society, Sandoz Ltd. (Basel), Wellcome Foundation, Wenner-Gren Foundation, W.H.O., and the delegates themselves.

We owe a great debt of gratitude to those who acted as session chairmen and section editors (listed in each volume), as well as to all the speakers. It was the responsibility of the chairmen to plan and conduct each session and to edit the proceedings, and it is they who have helped to ensure that publication can be rapid. I am especially grateful to Bill Lane-Petter, Edward Ford, Joe Herbert and Ken Joysey for the work they did in editing and collating the four volumes.

It was made clear during the Congress that events of the next few years will be critical for the survival of many primate species. And yet there are increasing signs of disunity among those most able to help them survive, with a corresponding delay in implementing the necessary education, habitat protection and research. Any tendency to polarise into "preservationists" and biomedical researchers must be resisted, so that there is united effort to help achieve the conservation of various ecosystems to the advantage of their occupants and the countries concerned. This is where "primatologists", with their unique spectrum of talents, should be able to play a leading role, and why regular meetings of an international society are important.

January 1977   D.J.C.

PREFACE TO VOLUME 2

The threat to conservation of primates comes chiefly from encroachment on, or destruction of, their natural habitats by man, but in some circumstances collection of live animals from the wild, whether for zoos or laboratory use, or their slaughter for food or as agricultural pests, may tip the balance significantly against ultimate survival. This is all the more true if wild groups are disrupted, or if live collection entails concomitant

Preface

killing of mates. Losses in the course of transportation to laboratory or zoo also lead to a magnified demand.

In this volume conservation is considered under three headings. First, a number of wild habitats are described, together with problems specific to them, including rehabilitation. Then come reports of methods of collection and transportation, and the way this traffic is conducted. Finally, the practicability of breeding primates in captivity is discussed; its technical aspects, its economics, and the ultimate effect it is likely to have on conservation.

W.L-P.

CONTENTS

List of Contributors

Preface

SECTION I   CURRENT PROBLEMS OF PRIMATE CONSERVATION   1
Chairman and Section Editor: R.W. Thorington (Washington, D.C.)

Conservation and primatology: a glum view of the future   3
   Richard W. Thorington, Jr.
The future of primate research   7
   G.H. Bourne
Current problems of primate conservation in Brasil   15
   P.E. Vanzolini
The status of primates in Guyana and ecological   27
   correlations for neotropical primates
   Nancy A. Muckenhirn and J.F. Eisenberg
Nature conservation in Indonesia and its problems with   31
   special reference to primates
   Edy Brotoisworo
Man's impact on the primates of Peninsular Malaysia   41
   Mohammed Khan bin Momin Khan
The conservation of nonhuman primates in India   47
   S.M. Mohnot
Primate conservation in Ghana   55
   E.O.A. Asibey
The conservation of primates in the United Republic of   75
   Cameroon
   Joseph Awunti
On habitat and home range in eastern gorillas in relation   81
   to conservation
   A.G. Goodall
Problems of primate conservation in a patchy environment   85
   along the lower Tana river, Kenya
   C. Marsh
Bioeconomic reasons for conserving tropical rainforests   87
   T.T. Struhsaker
The viewpoint of a conservationist
   R.S.R. Fitter

Contents

Summary remarks on primate conservation — 97
   Richard W. Thorington, Jr.

Round-table discussion on rehabilitation — 101
   Summarised by Monica Borner and Paul Gittins

SECTION II   TRADE AND SUPPLY OF PRIMATES — 107

Chairman and Section Editor: R.E. Hackett (Sussex, UK)

Introduction — 109
   R.E. Hackett

The supply of monkeys from Peninsular Malaysia — 111
   A.C. Laursen

The trapping and export of macaques in Indonesia — 115
   C.L. Darsono

Rhesus monkey supply and supply lines from trapper to user (quarantine facility) — 119
   D.A. Valerio

International traffic in primates from Thailand — 127
   Ardith A. Eudey

Transport of primates by air — 133
   G.E. Joss

Primate imports into the United Kingdom 1965-1975 — 137
   J.A. Burton

Economics and disease in imported simian primates — 147
   G.W. Tribe and D.A. Bassett

Cost analysis and rate setting in a primate animal laboratory — 153
   T.A. Fitzgerald

SECTION III   BREEDING PRIMATES IN CAPTIVITY — 163

Chairmen and Section Editors: W.R. Dukelow (Michigan) and W. Lane Petter (Cambridge)

Tupaia belangeri bias, an outbred stock of tree shrews — 165
   Anita Schwaier

The production of the common marmoset, *Callithrix jacchus* as a laboratory animal — 173
   W.A. Hiddleston

Breeding marmosets for medical research — 183
   S.F. Lunn and J.P. Hearn

An artificial milk for hand-rearing specified pathogen free marmosets, *Callithrix jacchus*, and the growth of animals on the preparation — 187
   J.A. Turton, K.R. Hobbs, D.J. Ford, J. Bleby and B.M. Hall

Reproduction in the squirrel monkey (*Saimiri sciureus*) — 195
   W. Richard Dukelow

Breeding *Alouatta caraya* in Centro Argentino de Primates — 201
   O. Colillas and J. Coppo

On the keeping, feeding and breeding of leaf monkeys in the Thüringer Zoopark of Erfurt — 215
   D. Altmann

Breeding baboons in Uganda and Cambridge — 217
   B.A. Baker, G.F. Morris and T.D. Cowan

Contents

The financial implications of breeding rhesus monkeys for biomedical research *R. Korte*	231
Reproduction in a closed colony of *Macaca arctoides* *J. Trollope*	243
Key Lois Island primate breeding project: a 5-year progress report *J. Pucak*	251
A programme of prepartum care for the rhesus monkey, *Macaca mulatta*: results of the first two years of study *C.J. Mahoney and S. Eisele*	265
Breeding statistics of *Macaca fascicularis* in Basel Zoo *W. Angst*	269
The influence of changed photic conditions on the breeding performance of laboratory primates (*Macaca mulatta*) *Frederick E. Birkner*	273
Breeding of macaques and chimpanzees at the Dutch primate centre *W. van Vreeswijk and H. Koning*	279
Adaptation of monkeys to extreme conditions *V.I. Chernyshev*	289
The economics of nonhuman primate conservation *L.B. Cummins, G.T. Moore and S.S. Kalter*	293
The breeding of nonhuman primates for biomedical research *J. Bleby*	297
Research management of a primate animal laboratory for medical experimentation by clinicians *J. Moor-Jankowski and E.I. Goldsmith*	301
Blood groups of apes and monkeys: their use and value in experimentation and breeding of primate animals *W.W. Socha, A.S. Wiener and J. Moor-Jankowski*	305
Buy or breed? *W. Lane-Petter*	309

SECTION I

CURRENT PROBLEMS OF PRIMATE CONSERVATION
(R.W. Thorington, ed.)

CONSERVATION AND PRIMATOLOGY: A GLUM VIEW OF THE FUTURE

RICHARD W. THORINGTON, JR.

*Curator of Mammals, Smithsonian Institution,
Washington, D.C. 20560 USA.*

The world will be very different when the International Primatological Society holds its biennial meeting in 2000 A.D. Judging from current trends, we can expect most, if not all, tropical forests to have been destroyed. The result will be that most of the primate species living today will be extinct in the wild. A few will survive in captivity, but not many for extended periods of time. These extinctions will have profound effects on our science. There was not a single discipline represented at the 1976 Congress which will be unaffected by mass extinctions of wild primates. The paleontologists will not be able to compare new fossil finds with recent forms as effectively as they now can. Medical scientists will find the number of species available as 'models' of human disease to be greatly restricted. 'Field' workers will be restricted to zoos and a few protected areas.

The conservation movement is attempting to slow or reverse the trends. To the extent that the conservation effort fails, our disciplines will be greatly impoverished. The future of the disciplines of natural history and ecology, in their broadest senses, depends especially on the conservation movement, particularly in tropical countries. This prompts me to present three different scenarios, successively more optimistic (from the viewpoint of conservation) and less probable than the one before, against which we can judge our present activities and their effectiveness.

1. The first scenario is that by the year 2000, all unprotected tropical forests will have been destroyed. Many of the areas which are now considered safe - reserves and national parks - will have been cut for wood, grazing and agriculture. The human population will have continued its logarithmic increase, and there will continue to be a strong demand for more goods per capita. This will place increasing and irresistible demands on the remaining reserves - demands that they be cut for wood or that they be put to more productive uses. Under these conditions, the earth's flora and fauna will be greatly impoverished. Whole tropical floras and faunas, including most primates, will disappear. The temperate zones and the arctic will be affected as well, through the demise of many migratory birds and mammals.

2. In the second scenario, there will be much less elimination of tropical faunas and floras by the year 2000. The parks and natural areas now established, and others established in the next decade, will still exist. Many primates and other animals will survive in these islands of natural vegetation within farmed and developed countryside. There will be continuing pressures to reduce the size and protection of parks, but these will be countered by strong and militant conservation groups. The result will be the conservation of some species from most of the major primate radiations, together with

their habitats.

3. In the third scenario, the human population will not increase as expected. The trend will be interrupted by calamities to man, in the form of war, economic collapse, or bad weather, all leading to mass starvations and deaths. Under such conditions it is possible that many animals would survive in natural habitats outside of parks as well as inside. Non-human primate populations would be little worse off or perhaps even better off than they are now.

The purpose of presenting the last scenario is two-fold. First is to emphasize how unlikely it is that the present situation is anything but a fleeting phase. The second reason is to suggest that the means of preventing great changes and many extinctions among primates are not available to us. Only great changes in the welfare of mankind will stop the current trends.

I believe that there are some important implications to the scenarios which I have presented. Since they are based on recognition of the increasing needs of increasing human populations, it seems evident that primate conservation may best be served by unconventional means. In fact, the most significant contributions to primate conservation may come from quite unexpected areas. It may be the person who develops an economical way to harvest wood from a mixed tropical forest in a way which favors the maintenance of the forest and its diversity. It may be the person who develops a more saleable veneer than that derived from tropical hardwoods. Or it may come from the person who demonstrates that there are more efficient ways to extract protein from a tropical forest than to cut it down, plant grass, and raise cattle.

A second implication of which I am convinced is that primate extinctions will be slowed significantly only by the conservation of the habitat of the animals. I am unconvinced that the existence or demise of the primate trade will greatly affect the outcome. We must emphasize habitat conservation if wildlife conservation is our aim. Similarly, we must realize that captive propagation in zoos and laboratories is not wildlife conservation. We shall probably be able to save some genetic stocks of primates in captivity. This is a worthy objective, which I strongly support, but it leads to the maintenance of domesticated species, not of wildlife.

If there is any validity to the views presented above, then it is obvious that we should try to influence events so that we are closer to the second scenario than the first, twenty-four years from now. This may seem like an unpromising prospect, but the International Primatological Society must remain dedicated to it. From an evolutionary perspective, these next two decades may be the most critical years of the whole Cenozoic for the order Primates. All of us must be alert to opportunities to ameliorate the predictable disaster. Just as natural history and ecology are dependent upon the success of conservation, so conservation is dependent on ecology and natural history. In one sense, conservation is applied ecology, and it can be only as effective as the basic science permits. The more we know about our fellow primates, the more likely it is that we can apply the proper biological strategies for their conservation.

There is a political side to the conservation movement. Once goals and objectives have been established, they must be accomplished within the political environment of the particular country and region involved. Ironically, very few members of the International Primatological Society live in countries with endemic primates. This is very unfortunate, because it is clear that primates will not be conserved just because some North Americans and Europeans with vested interests and peculiar perspectives have eloquently stated that they should be. We must involve our colleagues from countries which have wild primates. They are the ones who may have the opportunities to maintain the second scenario and to prevent the extinction of many species of primates. To this end the conservation symposia at the congresses of the International Primatological Society are more and more involving biologists and others from tropical countries. We must promote more exchange of ideas with these colleagues and we must encourage them to call on us for our help when they need it.

The present symposium was planned to include speakers from tropical countries of South America, Africa and Asia. Through the efforts of our British hosts, we were able to include six of these speakers. The following papers present their views on primate conservation in their respective countries. Their disparate views were greeted with animated discussion, some of which is incorporated into my concluding paper. Four other papers are recent syntheses or reports from primatologists on issues directly relevant to primate conservation. Between them they present a diversity of approaches and views on conservation problems of interest and relevance to the primatological community.

THE FUTURE OF PRIMATE RESEARCH

G.H. BOURNE

*Director, Yerkes Primate Research Center,
Emory University, Atlanta, Georgia USA.*

In the last 15 years there has been an outburst of research carried out on non-human primates. For many years previously they were imported for display in zoos and for use as pets. Their use in research reached a peak during the race to develop a polio vaccine.

At least in the United States, however, a steady decrease in the use of non-human primates for all purposes has occurred. This was evident as far back as the late 60's before the current limitations on the export of primates from their countries of origin were imposed. Kirton (1975) showed that the total imports of non-human primates numbered: 85,283 in 1968; 68,002 in 1969; 54,433 in 1970; and 48,660 in 1971. For 1974 and 1975 the figures were 51,253 and 36,202.

The present civilization is not the only one that was prodigal in its use of wild animals. In modern times a substantial number of primates were used to promote the health and welfare of mankind. However, the Romans in their games used vast quantities of all kinds of animals, purely for sadistic purposes. One set of games lasted 122 days during which 10,000 animals were killed. The opening of the Colosseum was celebrated by a 100-day show in which 4,000 domestic animals and 5,000 wild animals were killed. As many as 30 or 40 elephants would be killed in one series of games, hippopotamuses and rhinoceroses were killed in large numbers. The Romans exterminated the hippopotamus in the Egyptian Nile. Whole territories were entirely denuded of wild animals to supply the Roman arena. Several species were either exterminated or their numbers were reduced to the extent that later they became extinct, among these the European lion, the Libyan elephant and possibly the African bear. Apparently no apes were used in these games, but monkeys and baboons certainly were. Since the Romans, man has continued to be prodigal in his destruction of animal life. Some species in the past have also been virutally eliminated by the demands of fashion; one of the best examples of this is the Guereza monkey. In more recent times the factors that added immeasurably to the drain on wild primate populations were those of ignorance and indifference of the shippers of monkeys.

Take for example the area of Iquitos in Peru; 35,000 monkeys were exported yearly from there up until recently and 4 out of 5 were squirrel monkeys. Baranquilla and Leticia in Colombia were also outlets for live primates. There were always significant and often very large losses of monkeys during the capture procedure, and also during the holding and transit period, before the animals reached their final destination. In some cases the rate of loss has been assessed as being as low as 25%, but some estimates go as high as 80%. Since 35,000 live monkeys arrived in the United States from Iquitos alone, if the wastage is added on, this could raise the number of animals captured in this area to about 180,000 primates a year. It has been estimated that of every 200 squirrel monkeys captured in the Amazon basin for export to the United States, only 1 would be alive 12 months later. This applied particularly to animals for the pet trade. Near most settlements and most river highways in Peru and Colombia practically all primates have been eliminated. Hunters and trappers state that each year they have to go farther and farther into the forest to find an adequate supply of monkeys.

Although it is accepted that the eating of monkeys by the indigenous population is common, the extent to which this may occur is grossly underestimated. On September 25, 1976 an official of the flora and fauna section of the Agricultural Ministry of Peru reported that 7.5 million monkeys were eaten from 1964 to 1974 in Peru, against only 1.5 million killed or captured for export (Agence France Presse, 1975).

We know that monkeys and apes are widely used for food in Africa. Documentary films show pygmies shooting monkeys for food. Chimpanzees and gorillas are certainly killed for this purpose by other African tribes.

The development of modern weapons and the use of wire for trap making has made the capture and killing of primates all too easy and has increased their availability as food. The forest people of the African Coastal Rain Forest prefer monkeys as food to other game and while we do not have exact figures for the numbers killed for this purpose they must be considerable. The increase in human populations in these areas with their growing demand for protein makes the future of most primates in both the old and new worlds not very optimistic for this reason alone, nor will the establishment of forest reserves alone be the answer. In the years to come, political turmoil will continue to embroil many regions of the world which primates claim as their habitat.

Hartley (1972) has estimated that between 1954 and 1960, 1.5 million monkeys alone were used for the production and testing of polio vaccine. Much as we must deplore this prodigal use of monkeys we must also realize the benefits which they conferred on humanity in the numbers of lives saved and the thousands of children prevented from being permanently crippled. According to Hartley, in the United Kingdom in 1950 prior to the development of the vaccine there were 8,000 cases of polio of which 72% were paralytic. In 1969 there were only 9 paralytic cases. In the United States polio cases dropped from 40 per 100,000 of the population in 1952 to 4 per 100,000 in 1960. In 1964 the only polio epidemic in the United States was in the Yerkes Primate Center where a newly imported baby gorilla brought in the disease and communicated it to other gorillas and orang utans. One gorilla died during this epidemic and one gorilla and one orang utan were left partly paralyzed. This was the first evidence that great apes were susceptible to polio and it has led to the routine administration of polio vaccine to all infant apes at the Yerkes Center. Needless to say, we have had no further cases of polio.

Of the primates that were imported into the United States by no means all are for research; 50% went to zoos and petshops, more than half the remainder went to the pharmaceutical industry, much less than a quarter were actually used for research in universities or biomedical institutions. The United States has been currently importing 30,000 to 35,000 rhesus monkeys a year from India, but these have now been reduced to less than 15,000 by restrictions imposed by the Indian Government. Most of these animals were used by pharmaceutical companies and the rest were used for research.

The largest proportion of the monkeys coming out of South America have been squirrel monkeys and the majority of these have been going into the pet trade.

There is a Simian Society in the United States, with chapters in many parts of the country, and this society is formed of individuals who own pet monkeys and club together to learn more about their pets and how to look after them and feed them. Apart from these people the average monkey owner, with few exceptions, is totally ignorant of a monkey's needs both physically and emotionally and most of the monkeys that fall in their hands suffer an unhappy and painful existence and death. Apart from the danger to pet monkeys in this trade there is an even greater danger to the humans since most monkeys coming from the wild carry diseases often very dangerous to man and some South American monkeys have even been found to carry oncogenic viruses.

A lot of the research on primates is performed because they are the most desirable animals for particular projects, but some research workers have jumped on the primate bandwagon and are using non-human primates for experimental studies that could equally be performed on lower animals. Scott (1967) has rightly pointed out that sub-human primates are not just small human beings with fur coats, and there is no point in doing some studies on a primate just because it is a primate. They differ from each other as much as the various members of the carnivora do, e.g. bears, dogs, cats, raccoons. A primate should be selected for a particular piece of work only if it has a certain unique quality or qualities which make it an appropriate model for man and the objective of the experimentation is human health and welfare.

The term 'primate research' often is difficult to define, since there are so many different objectives in this type of research. Primates may be studied for their own sake. Such studies might include behaviour in the field or in the laboratory. Intelligence, neurology, cardiovascular system, endocrinology, general and reproductive physiology, as well as biochemistry, anatomy and pathology are areas of research on primates. In the past museums were great users of primates for taxonomic purposes, the objective being to get as many pelts and as much skeletal material as possible of a particular species to study variation within that species and accurately to delineate the species. For most smaller, very abundant primates, this would not have been an excessive drain, but with apes, it is another story. William Hornaday (1929), Chief Taxidermist to the United States National Museum, during an expedition to Borneo, killed 43 orangs on one day during his two years in the jungle. He must have made a serious dent in the orang population during his stay.

Although many of the primates coming from the wild into the United States have been going into the pet trade this has now been blocked by legislative action which became effective in late 1975 and is already resulting in a substantial saving in primate lives. In July 1976 the Canadian Government also placed a ban on the importation of primates except for research, educational and exhibition purposes. Of the others, those that are used for production and testing of vaccines and drugs are usually loosely described as being used in 'research'. There could be, however, a fairly large use of primates in the type of research that is needed to establish a new drug or compound of physiological importance such as a hormone. Djerassi (1970) estimated that the development of a new agent to control male fertility would require 500 primates just for testing its toxicity. To develop a luteolytic (abortifacient) agent would require 5,400 primates. This is the type of area in which increased use of primates might be expected.

If we are to have primates available in the future for study and to use for experimental and testing purposes we have to start introducing every form of conservation that we know of.

ANIMALS LEFT IN WILD

How serious is the situation among primates? Among the apes the condition of wild orang stocks rang the alarm through Dr. Barbara Harrisson (1963) some years ago. In the early 1960's she estimated only 3,000 orangs left alive in the wild. This was undoubtedly a significant underestimate, but it led to the establishment of many conservation measures which greatly reduced the drain on the wild populations. A Dutch biologist and his wife (Rijksen, 1975) working in Sumatra have recently estimated that there are 15,000 orangs left in Sumatra; there are probably at least 5,000 in Borneo so an estimate of 20,000 for the world population of orangs is a reasonable one. It should be noted, however, that the World Wildlife Fund, 1976, claims that only 5,000 orangs are left alive in Sumatra. Dr. Rijksen has pointed out that the habitat for half the Sumatran orangs is gradually being eliminated. The orang reservation in Sabah in North Borneo will remain intact so long as the Government continues to resist the pressures of the lumber firms which have their eyes on that choice piece of tropical forest located only 15 miles from a timber port. The tropical rain forest throughout the world is in fact being destroyed at a rate without precedent in human history. According to Thomas Lovejoy, program director of the World Wildlife Fund (United States), two-thirds of the South-East Asian rain-forests have already been eliminated, half of the African rain-forests are gone and even in the virgin forests of the Amazon, one-third has been cut over. According to Oppenheimer (1977), India, which at one time was covered almost completely with forest, now has only 17% of its area forested and much of this is poor quality forest, i.e. scrubby second growth.

The Eastern gorilla is suffering from the intense pressure for land use which accompanies population growth and, according to Goodall and Groves (1977), 'this is the ultimate creeping threat - and it is out of the hands of us all'. The Western gorilla and the chimpanzee are subjected to similar pressure as are many species of monkey. A case in point is the rhesus monkey. Most of us have thought of this species as very plentiful and we have certainly used them in a prodigal fashion as if they were. A million rhesus monkeys have been trapped and removed from India during the last 20 years. In the late 1950's 100,000 were exported a year, by the mid-1960's this had dropped to 50,000 per year and during the 1970's the number was a little more than 30,000 a year. Now the number available have been drastically reduced by the restrictions imposed by the Indian Government. Even with such a conservation measure Southwick and Siddiqui (1977) feel that unless there is more protection of rhesus monkeys by local people 'they will be eliminated within 25 years'.

There is no need to argue further the point before those already convinced that all forms of primates are in danger and that the danger is rapidly becoming more severe. If we are to continue to use primates for medical research and to continue to have primates available to study, positive action must be taken promptly. The thought that all we have to do is to establish primate reservations and police them, is well-meaning enough but impracticable if used as a sole method of conservation. No country can afford the level of policing that is necessary to maintain these reservations and they will remain viable only so long as the economic and political conditions permit. There has to be a multi-pronged approach to conservation. Endangered animals can be brought into areas where they can be more easily supervised and this need not even be in their country of origin. A good example is the proliferation of vervet monkeys at St. Kitt's Island, from a few introduced many years ago, where there are now said to be 10,000 of them. A similarly successful transplantation has been that of the rhesus monkeys to Cayo Santiago off Puerto Rico. If we are anxious to preserve a species we have to consider whether this preservation has to be in its native habitat and what we are preserving the species for. In some cases it may be easier to preserve a species in a part of the world not usually inhabited by them. It may even be necessary to preserve some endangered species in zoos or laboratories where they can be given top veterinary care. A good example of this is the Yerkes Center's colony of orang utans. Thirty-five orangs already living in the USA were acquired in the early 1960's. They were young animals at that time, they have now grown to maturity and, being given excellent nutrition and modern veterinary care, they have produced thirty-five live offspring. This colony provides a unique opportunity for the scientific study of the orang which is unequalled in the world and it is a situation which it will never be possible to duplicate. They have been used only for studies which do not risk either their breeding ability or their lives. The question arises whether we could not return some of these animals to the wild. Ideally this would be the thing to do if adequate funds were made available. However, there are two other considerations:

1. How long are the wilds to which the animals would be returned likely to remain intact?

2. Placing a captive bred animal back in the wild, even with training in a rehabilitation program, is a hazardous matter. Schaller noted a 23% mortality among baby gorillas born to wild groups he was observing and estimated that 40-50% of gorilla youngsters die before they reach six years of age.

The Yerkes Center is studying chimpanzees placed on an island off the coast of Georgia, 100 acres in extent and completely covered in wild vegetation. We have placed a number of chimpanzees on this island to form a viable social group. Although the animals' food is supplemented and they are carefully observed at their feeding station every day or two and sick animals pulled out for treatment, the mortality is high. This is another factor that must be taken into account in estimating the hazards to which animals in the wild are subjected, for infections and parasitism are a potent source of poor health. The expectation of life in the wild is as reduced for apes and other animals as it is for man in the wild.

Our gorilla colony, which numbers 18 animals, also has started active breeding. In the month of March 1976 we had three gorilla babies born to animals living in a one acre compound; all the mothers are successfully nursing their offspring. A total of 5 gorillas have been born to the Yerkes Colony in the last 4 years.

A criticism may be made of a laboratory breeding colony of orangs and gorillas that breeding may not continue into the second and third generation. If our orangs and gorillas follow our chimpanzees in this respect we will have no problem. We have two chimpanzees which are the great-great-grandchildren of two of the four chimpanzees which started the Yerkes Colony in 1930.

Chimpanzees are being rapidly depleted in the wild, but we have bred them successfully, and continue to breed them at the rate of 7 or 8 animals a year. In the 46 years since the Center has been established there have been 259 live chimpanzee births. Our total production of live baby apes up to July 1, 1976, has been 299.

From what has been said it becomes apparent that medical research with primates in the wild will become progressively more limited.

Among the factors which affect the numbers of primates are the following:
1. Consumption of primates as food.
2. Illegal poaching for a variety of purposes.
3. Loss of animals due to encroachment of man on their habitat.
4. Infection and parasitism in the wild.
5. Legal capture for (a) zoos, (b) pet trade, (c) vaccine production and testing, (d) medical research.

The numbers of primates collected for zoos, though considerable, does not approach those used as pets or for vaccine production and testing. There is a good chance that all our attempts to preserve endangered species of primates in the wild by the establishment of reservations will be unsuccessful. There is a case therefore, for selecting a few highly qualified zoos throughout the world to receive and breed endangered species. The same would apply to primate centers which in addition could make a scientific study of these animals.

In the area of vaccine production and testing there is a need that research will find methods of replacing primates. Human cell lines for example could ultimately replace monkey tissue for the production of polio vaccine. The use of *in vitro* models, e.g. tissue culture and chick embryos, instead of live models for a variety of studies and tests has barely been investigated. In the area of tissue culture itself the use of cell lines instead of primary cell cultures would be an important conservation measure in itself. In some areas the use of computers can reduce experimentation on live animals, but it is foolish to push this as if it can replace all live experimentation, as some people and organizations have tried to do.

Deliberate breeding of animals for vaccine work and for medical research is probably the best answer to the supply of primates for this purpose and there should also be breeding of endangered species to be sure that whatever happens in the wild these animals are preserved for posterity. What has already been done with a wide range of captive species is well described in the book "Breeding Endangered Species in Captivity" edited by R.D. Martin (1975). This book is by no means comprehensive, but it is a fine contribution to this area and demonstrates the importance of captive breeding in rare species conservation. As a matter of interest the Atlanta Zoo has had a spectacular success in breeding a rare species of crocodile (Morulet's) and has twice taken consignments of young crocodiles totalling 60 animals to Mexico to release them in their native habitat. Even if we do not release captive born endangered primates back to the wild, which would be a much more difficult thing to do, we could at least ensure the survival of the species that way. Consideration should be given also to establishing controlled breeding groups of endangered animals in their native countries, as the Zaire Government is trying to do with pygmy chimpanzees. This would probably be the best method of all of ensuring the survival of these animals. Consideration should also be given to the ethical considerations of breeding an endangered species in captivity for research purposes if it was a valuable model. Such a procedure would not endanger the species any further if only a small number were taken from the wild in the first place.

The current cutback in exports of rhesus monkeys from India and the restrictions on the export of monkeys from South America have stimulated programs to breed primates on a large scale. This will not only conserve wild primates, but since wild primates have many diseases it will produce a much better animal for research purposes.

The International Union for the Conservation of Nature is attempting to establish a set of guide lines for the use of non-human primates in medical research (Thorington, 1975). The list, drawn up by the Primate Specialist Group, IUCN and the International Primatological Society, is an excellent document and I believe could be supported by both the conservationist and the medical researcher if they operate via the cerebral cortex rather than the hypothalamus.

There is no doubt that every form of primate conservation is urgently needed and that international co-operation in this field was never more important. Fortunately, signs of this co-operation are already apparent. The United States, the United Kingdom, Indonesia, Zaire and some other countries in Africa and South America are already starting on co-operative programs which are most encouraging.

The preservation of our primate and indeed all of our animal resources is essential. Otherwise, the environment will become more and more sterile and all that we will be able to look forward to is a world shared by man and cockroaches.

REFERENCES

Agence France-Presse. (1975). Animal Welfare Institute Report, 24, No. 4, Washington, D.C.
Djerassi, C. (1970). *Science, N.Y.*, 169, 945.
Goodall, A. and Groves, C.P. (1977). *In* "Primate Conservation" (Rainier Prince de Monaco and G.H. Bourne, eds), Academic Press, New York (in press).
Hartley, E.G. (1972). *Brit. Vet. J.*, 128, 481.
Hornaday, W. (1929). "Two Years in the Jungle". Charles Scribner and Sons.
Kirton, K.T. (1975). *In* "Primate Utilization and Conservation" (G. Bermont and D.G. Lingberg, eds), John Wiley, New York.
Mannix, D.P. (1900). "Those about to Die". Panther Books, London.
Martin, R.D. (1975). ed. "Breeding Endangered Species in Captivity". Academic Press, London.
Oppenheimer, J.R. (1977). *In* "Primate Conservation" (Rainier Prince de Monaco and G.H. Bourne, eds), Academic Press, New York (in press).
Ryksen, H.D. (1975). Quoted in *International Zoo News*, 22, 132.
Schaller, G. (1963). "The Mountain Gorilla". University of Chicago Press.
Scott, J.P. (1967). *Ann. Rev. Psychol.*, 18, 68.
Southwick, C.H. and Siddiqui, F. (1977). *In* "Primate Conservation" (Rainier Prince de Monaco and G.H. Bourne, eds), Academic Press, New York (in press).
Thorington, R.W. (1975). Guidelines for the use of non-human primates in medical research. Submitted to IUCN General Assembly, September 1975.
World Wildlife Fund - Special Report (1976).

CURRENT PROBLEMS OF PRIMATE CONSERVATION IN BRASIL

P.E. VANZOLINI

*Museu de Zoologia, Universidade de Sao Paulo,
CP 7172, Sao Paulo, Brasil.*

INTRODUCTION

Our theme is the preservation of a vulnerable segment of the biota in a large and extremely diversified country at a critical stage of its history - the attempt to break the barrier of underdevelopment.
Conservation studies in South America have lacked concern for a theoretical background. An adequate, coherent body of ecological thought does not exist yet, but it is possible to place primate conservation within the framework of current evolutionary theory about the continental Neotropics, and to evaluate the resulting perspectives.
I shall develop my argument against the background of the doctrine of morphoclimatic domains, which seems to me the closest to the realities of South America, and which has proved extremely fruitful when applied to biogeographical and evolutionary problems. I call it a doctrine advisedly, because it is a line of thought, no more, from which pragmatic models can be extracted and tested against actual situations. It is still at a phase in which excesses are understandable and unavoidable. Indeed, these 'excesses' are desirable from the point of view of science; let natural selection prune the superfluous branches and strengthen the main line. Although I am sure the principles are valid for the tropics generally, I shall concentrate on the inter-tropical part of my own country, with which I am best acquainted and where I have seen several species of monkeys in the wild.

BACKGROUND

Morphoclimatic domains

A morphoclimatic domain, as defined by Ab'Saber (1967), is a macro-unit characterized by the superimposure of a set of features of relief, drainage, soil, climate, and vegetation. The continuous area where all the characteristic features occur together is called the core area. The core areas of the different domains are not juxtaposed (e.g. as biotic provinces would be) but are separated by intermediate belts of extreme diversity. Between two given core areas the same intermediate belt may have sectors of intergrading vegetation, of interdigitations and mosaics of the pure core vegetations, of peculiar types of vegetation not found elsewhere, etc.
It is essential to the evolutionary models based on this doctrine that these domains respond very quickly and drastically to climatic events - expanding, contracting, dissecting their area, actually pulsating during climatic

cycles. Evidence of these pulsations remains in the form of elements of one domain (soils, paleosoils, vegetational islands, relief shapes) left behind in the core area of a contrasting domain. These dynamic relationships between morphoclimatic domains, on a time scale of thousands of years, makes them more natural units for ecological, zoogeographical and, perforce, conservation studies than ecosystems. Inside and around a domain ecosystems interact, and the logic of the interaction is determined by the dynamics of the domains. It is a basic point of the doctrine that the ecogeographical relationships of the fauna with the morphoclimatic domains are all-important in explaining evolutionary patterns, especially at the species level.

Preservation policies

The ideal wildlife reserve would straddle at least two morphoclimatic domains and satisfy three lower constraints on size, i.e. be large enough (i) to accommodate the spatial inhomogeneities of the distribution of individual species, (ii) to prevent the extinction of the least dense species, especially the top predators and (iii) to ensure the survival of homeostatic mechanisms against irreversible oscillations of the systems involved. Additionally, its edges should be protected from degradation by a buffer zone.

These are, of course, necessary conditions to ensure the preservation not only of animal and plant species, but also of other environmental features as well as ecological and evolutionary processes. Theoretically, the 'ideal' reserve is one that would protect the most environmental diversity and active processes.

It is a truism to say that there are many problems in planning such a reserve. Some are geographical. For instance, the reserve must include adequate samples of the topographical features of the core areas and complete segments of the respective drainages (to the largest river). Intermediate belts between domains may present their own difficulties, especially because of their width, which may be of the same order of magnitude as those in the core areas.

Other questions are biological in nature. A very large number of animal and plant species are very heterogeneously distributed and considerable field exploration and mapping are needed to locate an optimal set of viable populations. It is necessary to determine the area needed for each of the rare species, and especially for the top predators (whose removal would start the whole system along new lines). The population parameters of such species are also required. Such determinations do not constitute too difficult problems but they are certainly time consuming and laborious.

When the geographical and biological parameters are known, it is certain that the ideal reserves will prove to be vast areas, perhaps only one order of magnitude smaller than a single domain. The need for homeostatic mechanisms imposes further demands. These are mostly related to extinction of local populations (e.g. due to local disasters or epidemics) and recolonization after such events. The conservation of these processes involves the preservation of larger, more diversified and dispersed populations, and thus the provision of more space. By the time everything is taken into account the size of reserve will have increased to the point at which the practical problems of acquisition and physical protection of the land become unmanageable. I doubt that an 'ideal' reserve will ever be established. The concept is important, however, as a standard against which to evaluate practical programs.

One first step down from the ideal would be reserves within the core areas of morphoclimatic domains. These would not protect the ecological and evolutionary processes occurring in the marginal and intermediate areas, but they would still permit proper conservation of biological diversity and of intra-system processes. The size and location of such reserves would still need to be determined with reference to the biological problems of inhomogeneity and rare species and would require reliable estimates of biological parameters and suitable safety margins. However, the reserves would be smaller (by at least one order of magnitude) and there would be a reduction in the practical difficulties consequent upon large size. Even so, these last remain so formidable that it is safe to say that very few reserves of this type will ever be established, and that their feasibility decreases steadily and rapidly with time.

A third approach would be opportunistic: the preservation, as possible and expedient, of individual species or sets of species, of peculiar landscapes, or of specific biological processes, or of combinations of these, in pristine or even in less-than-pristine conditions. In the real world, and especially in the real world of South American monkeys, this is the most practical course to take, if not already the only one left.

The word 'opportunistic' here has not the connotation of grabbing what happens to be available, but of making the most of the circumstances. Criteria of optimization are clearly called for, in the sense that as many as possible of the ideal features should be included. For instance, on the geographical side, typical landscapes of domain core areas might be chosen, as might well-defined sectors of transitional belts. If the emphasis is to be on the preservation of segments of the fauna, attention should obviously center on the species with inhomogeneous distributions, as their protection will automatically cover the evenly distributed forms.

The preservation of ongoing evolutionary processes has been usually neglected, but it is easy to see their importance and fragility. It is also easy to envisage the needed preservation. For instance, populations isolated from the main body of the species, especially when the length of isolation is known, are natural materials for the study of genetic and behavioral differentiation. Ecotones and other transitional areas are laboratories for the study of one of the least understood of all evolutionary phenomena (if it indeed exists): speciation without geographical isolation. Appropriate areas obviously need to be selected on the basis of evolutionary theory. They may present especially hard practical problems, stemming from the small size of the area to be protected, its awkward location, etc. This is perhaps the most interesting facet of the whole subject.

Such reserves will always be small and amputated segments of ecosystems and subject to many dangers (e.g. small population size, unbalanced predator-prey or competition relationships). This raises the issue of management, especially in those cases where the emphasis is on the preservation of species rather than processes.

Management is taken here as meaning strictly the avoidance of unwanted departures from a set of conditions considered desirable, or indispensable to the preservation of healthy stocks. One of the commonest types of disturbance in restricted spaces is disorderly population growth. Thus one is led to the idea of cropping, and next to that of exploitation. The management of relatively small reserves should always involve the study of population parameters relevant to the regulation of numbers, and to the eventual establishment of cropping programs for human utilization of the fauna. Such studies would also lead to the determination of optimal conditions for artificial or seminatural colonies.

I shall proceed now to apply these ideas to the actual situation in Brasil.

INTERTROPICAL MORPHOCLIMATIC DOMAINS OF BRASIL

Ab'Saber (1967), who established on firm grounds the concept of morphoclimatic domains and its application to Brasil, recognizes six domains; we need, however, to consider only the four intertropical ones:

The hylaea

Domain of the Amazonian forested lowlands; labyrinthic and/or meandric floodplains; extensive plateaux with semi-convex slopes; polyconvex low hills in crystalline areas; gravel and lateritic terraces; autochthonous black water rivers; perennial drainages.

The caatingas

Domain of the semi-arid depressions of the Northeast; covered by characteristic plant formations (caatingas); shallow decomposition of the bedrock; rock floors; extensive intermittent exorheic drainages; local braided channels; numerous fields of inselbergs.

The Atlantic forest

Domain of the forested 'seas of hills' of the Atlantic facade; with very deep regoliths; dense perennial drainages; characteristically polyconvex relief; meandric floodplains; extensive areas of superimposed soils.

The cerrados

Domain of the plateaux covered by 'cerrado' vegetation; penetrated by gallery forests; frequent extensive lateritic crusts protecting higher levels; sedimentary plateaux with gentle slopes; practically no polyconvex features of the relief; widely spaced drainage; headwaters in the shape of dales.

We prefer to refer to these areas by the names of the characteristic vegetation, for brevity and for mnemonic reasons, but the concept is an integrated one. Unless specifically stated, we are not dealing with the plant formations outside the core area of the domains.

THE HYLAEA

Paleoclimatic cycles and speciation. Evidence from many disciplines has been rapidly accumulating (latest comprehensive review in Journaux, 1975; see also Fairbridge, 1976) on the succession of relatively modern (on the scale of thousands of years) and pronounced climatic cycles in South America, but it is only in the area of the hylaea that their consequences on the distribution and differentiation of the fauna and flora have been studied in some detail.

We have just passed through a peak of humidity and are on our way to drier times, as indicated by the fact that the hylaea and the Atlantic forest, now separated by 1,300 km of open formations, still share many species that have differentiated only slightly (Vanzolini, 1974). The present patterns of distribution within the hylaea are thus the product of events that happened during and since the last dry period.

A number of speciation patterns of animals and plants that have been studied in some detail are compatible with the so-called 'refuge model' first proposed by Haffer (1969). This model postulates that the presently continuous forest areas were deeply dissected by inroads of open plant formations (cerrados and caatingas) during the dry episode. In the isolated forests thus formed both plants and animals underwent genetic differentiation. This is a simple and orthodox model, and a useful one. Many details
must be subjected to closer scrutiny, but nobody doubts that isolation in forest patches has led to differentiation. With the progress of the humid episode, these patches once more coalesced, and diverse distributional patterns arose. The most frequent is the presence of distinct taxa corresponding to the refuges, and meeting each other as subspecies or parapatric species. Occasionally, 'overriding' species may have taken over all or parts of the area, perhaps by 'swamping' or by competitive advantage.

It is to be expected that on the average the number of resulting taxa will be inversely proportional to the minimum area needed to maintain a viable population of the species, i.e. ultimately to the size and vagility of the animals: monkeys should show fewer successful isolates than butterflies.

This model suggests a conservation principle analogous to that derived from the concept of morphoclimatic domains: in the refuge areas we should find coadapted faunas evolved during isolation, while in the coalescence sutures we should meet ongoing intense processes of coadaptation. Thus, a reserve covering adequate adjacent portions of two coalesced refuges would be the optimal choice. The next choice (Thorington, 1975) would be the analogous one of a core area reserve, in this case a refuge area, being sampled.

Finally, the opportunistic attitude would be to identify areas of maximum density of inhomogeneously distributed species and preserve as much of them as feasible, including the relevant ecological and physiographical features.

Types of Amazonian forest, present status and perspectives. The Amazonian forest has three main facies: 'terra firme', never flooded, 'várzea', seasonally flooded, and 'igapó', permanently flooded. The total area of igapó is very small and unimportant to our argument. Várzea is said to occupy about 5% of Amazonia; in fact there is no reliable estimate, and the true value is probably much larger. Anyway, it is an enormously important component of the domain, being accessible, relatively fertile and adjacent to the resources of the rivers and lakes. Terra firme forest covers the interfluvial areas and reaches the rivers in spots. It is in general less fertile and not well provided with running water: the hylaean drainage is characterized by very large allochthonous rivers and 'ria' lakes with small and scattered tributaries.

The human occupation of Amazonia must be divided in two periods, before and after the introduction of earth-moving machinery.

The hylaea is usually considered impenetrable, aggressive and uninhabited: this is all myth. Where there is water there are people in the Amazon, and this has been so since pre-Columbian days, when large and culturally developed populations flourished in several parts of the valley (Meggers, 1971). Slash-and-burn agriculture has always been extensively practiced, both near the rivers and farther inland. After the colonization there have been extracting activities (rubber and other gums, Brasil nuts and other fruits, fibers, skins, etc.) literally all over the domain. There is no extensive tract of really 'virgin' forest in Amazonia, and very little that has not been exploited in some fashion by neo-Brasilians. And in most areas the return of the forest after even small scale devastation is reluctant and imperfect.

Figure 1. *Amazonian highways against the backdrop of the morphoclimatic domains.*

The degree of damage done varies, of course, from area to area. Along the main river and the large tributaries the disturbances are extensive, marked and apparently irrevocable. Between rivers, damage appears minimal - although we have no actual standards for comparison. It seems that survival-level hunting and gathering activities can be compensated by natural homeostatic mechanisms.

The practical result from our present viewpoint is that until fairly recently large areas between the major tributaries could be considered available for the establishment of 'virgin' reserves, and there was no real hurry to get things done.

The tractor has changed all this. The seemingly impenetrable forest crumbles under the blades: the superficial root systems offer no resistance to the machine, and the work of road opening and land clearing is very easy. Crossing from river to river and clearing the interfluvia has become an everyday task. In consequence the amount of available land for reserves is shrinking faster and faster, and some paleoclimatic refuge areas are already gone

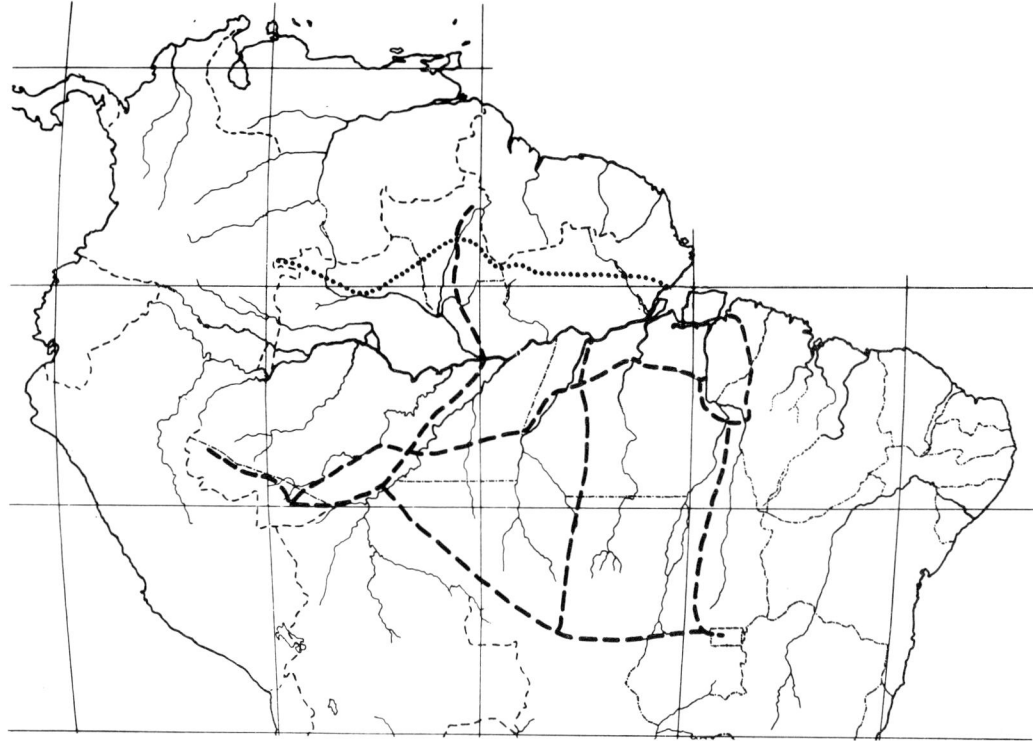

Figure 2. *Amazonian highways against the backdrop of the drainage.*

(eastern Pará is an example) or will disappear soon.

This technological change came about at a time when the psychological need to 'open' and 'integrate' Amazonia was at a peak, and resulted in much road-building (of a strategic nature) and large scale colonization, encouraged by tax deductions.

Figures 1 and 2 show the network of roads built or under construction, against the background of the morphoclimatic domains and of the drainage.

The 'Transamazônica' is practically completed now. It is an extremely well designed strategic road, connecting the lowermost rapids of all the major tributaries, and thus linking trucking to fluvial transportation. Colonization along the road has not been successful. Only three areas are being actively settled, and of these only one intensively so; here the belt of damage is some 100 km wide. Of course, since the road permits steady penetration, it is to be expected that this influence will become progressively more marked.

Of the three roads from the South, two (Belém-Brasília and Porto Velho-Manaus) are not only open, but already paved. The third, and central one (Cuiabá-Santarém) is under construction. Judging from what happened along the oldest of the roads (Belém-Brasília), each one of them will be bordered by a deforested belt some tens of kilometers wide.

The northern perimeter road is being opened at a slower pace. It is not expected that the road itself will cause much damage, but again it will provide access.

The gravest threat to the integrity of the hylaea is at present the clearing of forest for cattle raising. This is considered an admirable goal by a majority of the people with power of decision in the 'opening' of Amazonia.

The southern edge of the domain and the intermediate belt with the cerrado are being actively attacked now, from the westernmost state of Acre, where there has been a recent land boom, to southern Pará, where large companies primarily involved in other pursuits (e.g. automobile manufacturing or television programs) are investing their specific tax deductions. Extant forestry laws, not entirely inadequate in concept, are not being observed. Although the actual amount of deforestation is not accurately known it is certain to be extensive.

Logging is not very actively pursued yet. The extreme diversity of the forest and the difficulty in working the swampy areas and the interfluvia have so far been effective deterrents. Again the introduction of the tractor is changing the picture, and there is notice of plans for logging large areas and (to comply with forestry laws) plant them with crop trees (palms, cocoa, etc.). The disastrous results of rubber plantations in the past make it unnecessary to dwell on the dangers inherent to large areas of uniform vegetation in the tropics.

Finally, mining must be considered. Some 85% of the sediments carried by Amazonian rivers are allochthonous; they are deposited as the rivers enter the plain and lose speed. In consequence there has always been a certain amount of placer mining in Amazonia. This has not been an ecological threat until the recent introduction of heavy machinery. We are facing now the first attempts at strip mining.

Perspectives for conservation. As I write I can still think of a few areas, principally north of the river, where large samples of hylaea, of the order of tens of thousands of square kilometers, including very characteristic areas, if not actual paleoclimatic refuges, can still be preserved. The two major practical difficulties are the fact that much of the land is owned and must be bought, and that human populations have to be moved. These problems are not too serious. Land away from roads is still cheap (a few dollars per hectare) and the people have a semi-nomadic tradition and would be amenable to displacement. Worse are the difficulties in planning due to the lack of ecological surveys, and the ever present danger of the discovery of minerals. Of course, a program of this magnitude could exist only as part of a national plan of conservation, which does not exist now and is not contemplated in the near future.

On the opportunistic side there are myriads of alternatives. An example is the ongoing project of an 'ecological station' in the archipelago of Anavilhanas, a labyrinth of 'new land' deposited by the Rio Negro some 70 km upstream from Manaus, where the interface between water and land in an alluvial valley can be marvelously studied.

From the viewpoint of primatology, there are still many areas, on both sides of the river, where reserves of hundreds of square kilometers, permitting the maintenance of a reasonable degree of biological diversity, can be established and operated at no high cost. Time for many of those, however, is running short, and there is no program for surveying the flora and fauna and planning a network of reserves.

Finally, a potentially very rewarding field is that of small reserves, dedicated to a single or a few species. It is known (Thorington, 1969) that many primates will thrive under such circumstances, and the very need for managing the areas by cropping will certainly relieve the pressure on the populations under more natural conditions. The cost of land would be larger than for the other types of reserves, but the small size and ease of maintenance would materially lessen the burden.

THE ATLANTIC FOREST

The great and legitimate preoccupation with Amazonia has somehow pushed the Atlantic forest into the background. It must be remembered, however, that its small primate fauna is of great importance and that conservation problems there are the gravest of all.

The lack of relationships between the primate faunas of the Atlantic forest and the hylaea is really remarkable, especially when compared with other vertebrate groups (Haffer, 1974; Vanzolini, 1974). The only species in common is the ubiquitous *Cebus apella*; *Brachyteles* and *Leontopithecus* are endemic monotypic genera; *Callicebus personatus* and *Alouatta fusca* are endemic species. All these forms pose important evolutionary problems, but only *Leontopithecus* and *Brachyteles* have been the object of some study and protection (Coimbra-Filho, 1972; Aguirre, 1971). The howler and the titi are not even mentioned by Coimbra (loc. cit.), in spite of being obviously as endangered as any other Atlantic forest mammal.

In fact, the Atlantic forest is mostly gone, and it remains under heavy predation. It extended originally between 7 and 29 degrees of latitude south, and constituted the facade of Brasil. Until very recently its fertile soils maintained the growth of the country, and it still supports a major share of the population. The remaining forested areas (apart from a few and not well kept reserves), along the coast and on the coastal mountain range of southeast, are now under the new threat of recreation needs, fostered by new roads. There is very little prospect of any coherent and extensive conservation program; salvage is all that remains. The problems of expensive land and of opposition from persons with vested interests are very serious, and are compounded by our lack of recent information on the status of the species. I would consider studies of *Callicebus personatus* and *Alouatta fusca*, leading to their protection, to be urgent tasks of Brasilian primatology.

CERRADOS AND CAATINGAS

The enormous diagonal belt of open formations that stretches from the coast of northeastern Brasil to northern Argentina houses the only South American primates adapted to open formations: *Alouatta caraya* and *Callithrix jacchus*.

Speciation in contrasting and adjacent environments is not an infrequent phenomenon among South American vertebrates, but it is one for which no convincing mechanism has been proposed. Even Haffer (1974) in his thoughtful review of the toucans, dismisses *Ramphastos toco* with only the comment that it is 'ecologically aberrant'. The problem is a fascinating one from many viewpoints, such as the genetic mechanisms and geographical configurations that can permit the necessary reduction of gene flow for full speciation, the preadaptations towards the great physiological and behavioral changes involved etc. I would consider these two species among those most needing study and protection.

There has been an enormous development of the cattle industry in Brasil in recent years. Actual figures on the areas involved are hard to obtain, but informed opinion suggests that it is already extremely difficult, if not impossible, to establish a cerrado reserve of domain size, especially one encompassing a good area of Amazonian buffer forest as is sorely needed. Much of the land is owned by strong owners, and the general attitude in the country is for more and more beef. Recent troubles experienced by Indian tribes under government protection indicate the vigour of this spirit.

A few reserves, one order of magnitude below domain size, are still feasible, but there is not much choice as to where to locate them. They should afford adequate protection to practically all the fauna. On the other hand, their establishment, again, must be part of a national plan of conservation.

The situation in the caatingas is sui generis. This is economically the poorest region of Brasil, suffering from all the drawbacks that poverty entails. At the same time it is the home of the hardiest, most energetic and enterprising breed of Brasilians, who have extracted maximum yields from the environment for many generations. The second growth vegetation closely resembles the original one physiognomically; thus it is hard to discern how disturbed the domain is, and how the disturbances are distributed geographically. I know of no estimates. Given the density of human occupation, it seems out of the question to establish very large reserves encompassing the diversity of the area, which is still not very well understood. A network of small reserves seems, at present, the more reasonable approach, and one that would ensure adequate protection to a sizable segment of the fauna.

CONCLUSION

The framing of a conservation policy depends on the firm definition of national aims and on the acquisition of basic scientific information. Both things are lacking in Brasil.

The state of underdevelopment is an anxious one. In yearning to escape oppression and to overtake the industrialized countries, the underdeveloped countries ape concepts, attitudes and practices - with an unfortunate time lag and within an incongruous technological context. Even when modern concepts (such as the need for conservation) are adopted, it is at a very shallow level. The means to protect half a dozen marmosets must be obtained from people who honestly believe that heavy city pollution is a fair price to pay for industrialization. Conservation as a national goal may receive some lip service, but scarcely anything else.

The dearth of scientific information is glaring. The scientific establishment is small and largely incompetent, from the level of basic research to that of civil service expertise. The acquisition, digestion and application of biological data is slow, haphazard and unsustained.

In brief, a general solution for the problem of primate conservation in Brasil will ultimately be part of a general solution for the problem of poverty and ignorance, i.e. of underdevelopment. This uphill struggle is slow, and the pace of technological destruction is fast. Missionary work of enlightenment should of course not stop, but the most profitable course at present is to concentrate on medium and small scale projects capable of short term completion.

REFERENCES

Ab'Saber, A.N. (1967). *Orientação*, (Dept. Geogr. Univ. S. Paulo). 3, 45-48.
Aguirre, A.C. (1971). "O mono Brachyteles arachnoides (E. Geoffroy)". Academia Brasileira de Ciências, Rio de Janeiro.
Coimbra-Filho, A.F. (1972). "Espécies da Fauna Brasileira Ameaçadas de Extinção". Academia Brasileira de Ciencias, Rio de Janeiro.
Fairbridge, R.W. (1976). *Science, New York*, 191(4225), 353-359.
Haffer, J. (1969). *Science, New York*, 165(3889), 131-137.
Haffer, J. (1974). *Publ. Nuttall Ornith. Club*, 14.

Journaux, A. (1975). Recherches géomorphologiques en Amazonie brésilienne. *Bull. Centre Géomorph. C.N.R.S. Caen*, 20, 1-67.

Meggers, B.J. (1971). "Amazonia: Man and Culture in a Counterfeit Paradise". Aldine-Atherton, Chicago and New York.

Thorington, R.W. (1969). *An. Acad. Brasil. Ci.*, 41 (Suppl.), 253-260.

Thorington, R.W. (1975). *In* "Symposia of the Fifth Congress of the International Primatological Society" (S. Kondo, M. Kawai, A. Ehara and S. Kawamura, eds), pp. 547-553. Japan Science Press, Tokyo.

Vanzolini, P.E. (1974). *Papéis Avulsos Zool., S. Paulo*, 28(4), 61-90.

NOTE ADDED IN PROOF

Since this paper was written the Brasilian government published an important document on conservation in Amazonia: Wetterberg, G.B., M.T.J. Padua, C.S. Castro and J.M.C. Vasconcelos, 1976. Uma análise de prioridades em conservação da natureza na Amazônia. 62 pp. PNUD/FAO/IBDF, BRA-45, Série Técnica No. 8. 62 pp.

THE STATUS OF PRIMATES IN GUYANA AND ECOLOGICAL CORRELATIONS FOR NEOTROPICAL PRIMATES

NANCY A. MUCKENHIRN* AND J.F. EISENBERG**

*ILAR-NRC, 2101 Constitution Avenue, Washington, D.C. 20418, USA
**National Zoological Park, Washington, D.C. 20009, USA

A field team conducted a three-month survey of the lowland rain-forests of Guyana during 1975 under the auspices of the US National Academy of Sciences and through the co-operation of the Pan American Health Organization (Muckenhirn et al., 1975). The team observed the same eight primate species in Guyana that Husson (1957) recorded in Surinam. Four species were observed on both sides of the Essequibo River. The other four were observed only to the east of the river and not in the northwestern section of the country. If the Essequibo River drainage acts as a distributional boundary for these species, then the single national park which lies on the northern bank of the Potaro River at Kaieteur Falls will offer protection for only half of the endemic species in Guyana.

TABLE I

Factors Contributing to the Status of Primate Populations in Guyana

Species	Density Rank	Distribution in Guyana[a]	Reproductive Rate[b]	Sensitivity to Human Encroachment[c]	Proposed Status[d]
Saimiri sciureus	1	++	++	-	1
Cebus apella	2	+	++	--	2
Alouatta seniculus	3	++	+	----	3
Cebus nigrivittatus	4	++	++	--	2
Chiropotes satanus	5	+	++	---	3
Pithecia pithecia	6	++	++	---	3
Saguinus midas	7	+	+++	-	2
Ateles paniscus	8	+	+	----	4

[a] + = restricted distribution, ++ = widespread distribution.
[b] + = low rate, ++ = moderate rate, +++ = high rate.
[c] - = highly adaptable, -- = moderately adaptable, --- = slightly adaptable, ---- = not adaptable.
[d] Vulnerability is ranked from 1 (least vulnerable) to 4 (most vulnerable). See text.

The eight species of primates are listed in Table I. We attempted to develop a measure of the relative status of the different species by looking at four characteristics: density, distribution, reproductive rate, and susceptibility to human encroachment. We first ranked the species by their overall population densities, based on censuses conducted on transects along both trails and rivers (Table I). Second, we distinguished two distribution patterns; four species apparently occur throughout Guyana and the other four species appear to be absent in the northwest. Third, we ranked the reproductive rates of the eight species; the ability of a species to maintain its numbers against natural losses and removal by humans is related to its reproductive rate. *Saguinus*, with its ability to produce more than one young per year and to mature in less than two years has the highest reproductive potential among South American primates. The large-bodied *Alouatta* and *Ateles*, have the slowest replacement rate among cebids, with their long inter-birth intervals of more than one year and pubertal age exceeding three years.

Fourth, the estimate of sensitivity to human encroachment was subjective and combined such factors as diet, adaptability to an environment modified by human activity, and susceptibility to hunting. The four species that appear to be highly to moderately adaptable are those that thrive in mixed habitats and consume insects as well as fruit. Because considerable riverine and regenerating forest is associated with slash-and-burn agriculture, the squirrel monkey, two capuchin species and tamarins seem to increase initially in areas of expanding agriculture. In fact, *C. apella* is considered to be an agricultural pest in areas of developing corn production.

The four more sensitive species appear to have more stringent habitat requirements. Forests with relatively interlocking canopies are necessary to offer continuous arboreal travel routes for *Alouatta* and *Ateles* which have prehensile tails. Little is known of the natural history of the pithecines, but *Chiropotes* appears to be a fruit feeder of the upper canopies and *Pithecia* appears to be a generally uncommon monkey of the lower and middle strata.

By combining these four factors, we rated the species on a one-to-four point scale of increasing vulnerability. Squirrel monkeys appear to be most able to maintain populations in the face of expanding agricultural land use while *Ateles* appear to be the most vulnerable. Conservation measures will be needed if the spider monkeys are to continue to be widely distributed throughout the country as they are today.

The survey results from Guyana are similar to those of Bolivia and Peru (Table II). When we excluded the extreme sample sites in which either a single species was observed or the density estimates were exceptionally high, we found that 60% of the survey sites in Guyana had 2-6 species, 2-11 groups/km^2 and 12-67 individuals/km^2. These differ only slightly from the Peruvian and Bolivian data. We suggest that the central values represent conservative but typical survey results in lowland forests using transect methods. We combined the data from the three countries to form the hypothesis for future surveys that 50-60% of the areas will have 2-7 species/site, 2-11 groups/km^2 and 8-80 individuals/km^2.

Several correlations between the presence and the relative abundance of different species and the carrying capacity of an area have become apparent from recent studies. Five of the correlations are given below.

1. The density of *Aloutta* is lower in regenerating forests or second growth areas and comparatively high in mid- and late-successional areas. Data accumulated over thirty years on Barro Colorado Island, Canal Zone (Carpenter, 1964; Chivers, 1969) support this correlation as do locality comparisons in Guyana and Venezuela (Eisenberg, 1977).

2. Species diversity of primates is highest in areas where there is a good reason to believe that the carrying capacity is high, i.e. the overall biomass

TABLE II
Range of Primate Estimates from Transect Surveys

Guyana Trail and River Surveys
(Muckenhirn et al., 1975)

	Low Estimates	Central Estimates	High Estimates
No. sites (15)	3(20%)	9(60%)	3(20%)
Species/site	1	2-6	2-4
Groups/km^2	< 1-2	2-11	20-41
Individuals/km^2	2-10	12-67	138-812

Peru and Bolivia Trail Surveys
(Freese, 1975; Heltne et al., 1975)

	Low	Central	High
No. sites (11)	1(9%)	6(55%)	4(36%)
Species/site	1	3-7	4-8
Groups/km^2	< 1	2-10	15-23
Individuals/km^2	5	8-80	108-287

Overlap in Studies

	Low	Central	High
No. sites (26)	4(15%)	15(58%)	7(27%)
Species/site	1	2-7	2-8
Groups/km^2	< 1-2	2-11	15-41
Individuals/km^2	2-10	8-80	108-812

exceeds 100 kg/km^2. The greatest diversity of nine species at a single site has been reported from Colombia (Izawa, 1967).

3. Inter-taxa groups associate at sites with high carrying capacities. Examples include mixed-species groups of *Cebus* and *Saimiri* in Guyana and Peru (Freese, 1975) and *Saguinus fuscicolis* and *S. imperator* also in Peru.

4. Areas of reduced overall primate densities have reduced numbers of larger cebids (*Aloutta*, *Ateles*), but pithecines are relatively more abundant in such areas. These forests are characterized by white sands and primarily black water drainage systems or are distant from rivers. Examples include the relatively high numbers of *Pithecia pithecia* in the HMPS Forest Reserve and in the clump wallaba forests (*Dicymbe* spp.) observed along the Issana Road in Guyana.

5. In areas of reduced carrying capacity there is also evidence of competition between taxa and ecological dominance by one species over another. Examples include areas where *Saimiri sciureus* and *Saguinus midas*, *Chiropotes satanus* and *Pithecia pithecia*, *Cebus apella* and *C. nigrivittatus* overlap in Guyana. In such habitats, the former species occurs at higher densities than the latter species.

REFERENCES

Carpenter, C.R. (1964). ed. "Naturalistic Behaviour of Non-human Primates". Pennsylvania State University Press, University Park.

Chivers, D.J. (1969). *Folia Primatol.*, 10, 48-102.

Eisenberg, J.F. (1977). Habitat Economy and Society: Correlations and Hypotheses for the Neotropical Primates. Paper presented Burg Wartenstein Symposium No. 75 (I. Berstein, organizer), 12-21 August 1977.

Freese, C. (1975). *In* "Primate Censusing Studies in Peru and Colombia". Pan American Health Organization, Washington, D.C.

Heltne, P., Freese, C. and Whitesides, G. (1975). "A Field Survey of Non-human Primate Populations in Bolivia". Pan American Health Organization, Washington, D.C.

Husson, A.M. (1957). *Studies on the Fauna of Suriname and other Guyanas*, 2, 13-40.

Muckenhirn, N.A., Mortensen, S., Vessey, S., Fraser, C.E.O. and Singh, B. (1975). Report on a primate survey in Guyana. Pan American Health Organization, Washington, D.C.

Izawa, K. (1977). Primates, 18, in press.

NATURE CONSERVATION IN INDONESIA AND ITS PROBLEMS
WITH SPECIAL REFERENCE TO PRIMATES

EDY BROTOISWORO

*Faculty of Biology, Gadjah Mada University,
Yogyakarta, Indonesia.*

INTRODUCTION

Indonesia covers thousands of islands between the mainlands of Asia and Australia. Many different habitats occur there, and the variety of plant and animal species is enormous. Several endemic species occur, the result of thousands of years of geographical isolation; among these are several primate species. Thirty primate species in all, excluding *Tupaia*, are found in Indonesia; they occur from Sumatra in the west to Sulawesi (Celebes) and Timor in the east. In the past, most of the animal species lived together in harmony with 'primitive' man. Technological progress, which, of course, cannot be avoided, combined with a rapidly growing population, upset the harmonious balance between environment and man. In some places in Indonesia we can still find people living in harmony with their environment, as on Siberut Island (Mentawai Islands), in central Sumatra (the Kubu) and in central Kalimantan (the Punan). Such relicts, however, are usually considered 'primitive'. The imbalance nowadays results in ecological problems which are aggravated by unwise commercial logging practices started some years ago.

In this paper several of the problems concerning nature conservation will be discussed, with emphasis on the primate conservation of the islands of Java, Bali, Sumatra and Mentawai on which the author's experience is mainly based.

NATURE CONSERVATION IN INDONESIA

Conservation of nature in Indonesia was started in the 'Dutch time' by the decree of law and acts in which hunting and other activities were regulated. Seventeen laws and regulations were issued between 1909 and 1941, some of which are still valid (Basjarudin, 1976). Modernising legislation dealing with management and conservation is still in process and will hopefully come into effect soon.

According to the official record of the Directorate of the Nature Conservation and Wildlife Management, as of May 1976 there were 176 established reserves covering 33,023 km^2; 38 reserves are still in the process of being completed and 32 other reserves have been proposed (Table I). The Government plans ultimately to enlarge the area to 100,000 km^2 approximately; by the end of the Second Five-Year Development Plan (i.e. in 1979) it is intended to increase the reserved areas by approximately 50,000 km^2. The number of protected species will increase accordingly. At the moment some 355 animal species (belonging to 51 families and about 210 genera) are protected by law (Pranowo,

TABLE I
The Distribution of Reserves by Island

Island	Established	In Process	Proposed
Sumatra	39	11	3
Java	94	8	8
Nusa Tenggara (Lesser Sunda Islands)	10	2	2
Kalimantan (Borneo)	12	7	3
Sulawesi (Celebes)	13	5	14
Maluku (Moluccas)	7	3	-
Irian Jaya (West New Guinea)	1	2	2

1975). This number reflects praiseworthy efforts on the part of the Indonesian nation to protect their resources from extinction, saving them for the benefit of mankind as well as science. Such 'paper protection' is, however, unaccompanied by managerial improvement to overcome the present problems which threaten the safety and perpetuity of many important reserves.

Only recently has nature conservation in Indonesia become effective. In the past an inadequate number of officials were appointed to manage and guard the reserves. The governmental budget was insufficient to pay for proper management of the reserves. In addition most employees had little education in the conservation of nature and its problems. Therefore hunting, animal capture and forest destruction prevailed almost everywhere. The consequences of mismanagement or lack of management can readily be found in Indonesia; for instance, several species are rapidly decreasing in numbers, or are in danger of extinction. Many protected species are still kept as pets. Several primate and bird species are clear examples of this. Also there is the extensive trade in stuffed animals. Recently (since about 1970) conservation of nature in Indonesia has seen some major improvements; international aid, such as that provided by the World Wildlife Fund, played a major role. The governmental budget has increased and the number of officials in the Nature Conservation service has risen. These improvements are still not adequate, however, for managing the entire reserved area properly. Scientific interest in the numerous biological wonders of the tropical rain-forest within Indonesia and abroad is increasing, but there are still many problems. Some problems are caused by well-educated people who may know the nature conservation laws, but yet are not concerned about them since there is inadequate law enforcement of the nature conservation laws and acts. Some people even feel that they are privileged and can ignore the law. If the government does not take measures to halt this and does not enforce the nature conservation laws, there is imminent danger that more people will eventually ignore the law.

Indonesia is an enormous country, covering thousands of islands. Distance fosters disagreements and differing attitudes between the Nature Conservation service and the Forestry service at local service levels (Rayon level). At the higher levels both Directorates work co-operatively. The different attitudes are caused by conflicts between conservation and utilization. The forestry service often does its work by neglecting nature reserves or nature conservation acts because of its more immediate commercial interest. The nature conservation service has to guard future resources and thus cannot deal with such immediate commercialism; this happens, for instance, in some places in Sumatra and Java. Consequently strain emerges between the two services, which has a very detrimental effect on conservation. Actually the conflict would not arise if there was mutual understanding and realization that they have the same final goal and objectives, as declared in the Fifth World Forestry Congress (Daryadi, 1976): '... management of forest in a manner that will conserve the basic land resources while yielding a high level of production in the five major uses - wood, water, forage, recreation and wildlife - for the benefit of the greatest number of people in the long run'.

RAIN-FOREST AND TIMBER CONCESSIONS

In Indonesia, sixty percent of the total land area is forest (Soedjarwo, 1975). At present the Indonesian government allocates large areas of forest for commercial logging. Due to exportation of wood, forestry is second only to the oil industry in yielding foreign exchange.

Forest in Indonesia can be classified in three categories, namely *protected forests* (including reserves), *production forests* and *reserved forests*. Protected forests are forested areas maintained to protect the land and to prevent floods, erosion, etc. This forest is situated mostly above 500-metre elevation, or on very steep terrain. A production forest is a forested area designated for wood production. A reserved forest is a forest which can be converted into various uses or which can be kept undisturbed before any further utilization. Protected forests cover about 570,000 km^2 whereas there are about 380,000 km^2 of production forests. This is about 28% and 18% of the total landmass respectively. The areas designated for nature reserves total 5-10% of the forest area. Table II shows the forest classification in Indonesia.

TABLE II
Forest classification in Indonesia

Type of Forest	Conditions				Total
	1	2	3	4	
Protection Forest and Reserves	550,000	-	-	20,000	570,000
Production Forest	-	160,000	40,000	180,000	380,000
Reserved Forest	-	130,000	20,000	100,000	250,000
Total:	550,000	290,000	60,000	300,000	1,200,000 km^2

Cited from Penjajagan Reboisasi Nasional di Indonesia, published by the Direktorat Reboisasi dan Rehabilitasi (1971).
Note: 1, undisturbed; 2, productive; 3, of low productivity; 4, damaged.

As shown, the total forest area in Indonesia should be 1,200,000 km^2. 380,000 km^2 are set aside for production forest; 250,000 km^2 are reserved; so 630,000 km^2 can be used for timbering. Several million hectares suffered heavy damage and became totally unproductive, mostly covered with lalang (*Imperata cylindrica*). Kartawinata (1975b) reported 160,000 km^2 occupied by lalang and 230,000 km^2 covered with secondary growth. The protected forest which is still largely undisturbed amounts to 550,000 km^2.

Until March 1974, 180 different timber companies were registered by the Directorate General of Forestry to operate in Indonesia. They utilized an area of 173,000 km^2. In addition, 88 different companies were approved to go into operation (71,000 km^2 of forest) and 374 other companies have submitted proposals to operate 362,000 km^2 (Hamzah, 1975). Thus the total forest land which will be logged is to be about 606,000 km^2.

By the regulation of the Directorate General of Forestry No. 35/Kpts/DD/I/1972, there are three methods by which wood can be exploited from the forest, namely:

1. Selective logging, a method that is considered the least destructive from an ecological viewpoint, since the damage inflicted to the forest is not as detrimental as the latter methods. By this method, only trees more than 50 cm across at breast height can be removed and the cutting rotation is 35 years.

2. Clear-cutting with natural regeneration. This is practicable if the number of seedlings of certain tree species is sufficient for natural regeneration. This could be proved by Linear Sampling Milliacre.

3. Clear-cutting with artificial regeneration, a method that resembles the second except that regeneration is accomplished by planting. The cutting rotation of the latter two methods is 70 years. Actually the regulations concerning cutting methods are very strict and always require the least possible damage to the forest ecosystem. However, control is inadequate during the actual field operations of the commercial logging companies and regulations are often violated. As the companies seek as much profit as possible, smaller trees are removed and the damage to the forest can better be described as destruction (see Burgess, 1971; Meyer, 1973). Moreover, the status of several reserved areas is such that logging practice can be permitted and this results in a considerable degraded habitat. When reserves are meant to preserve the intact and original habitat for future generations, no commercial logging should be permitted as it is incompatible with conservation.

Little is known about the influence of logging on the forest habitat and the animal species living therein, but it is clear that any commercial logging activity seriously impoverishes the species composition (Burgess, 1971) and dramatically disrupts the forest structure (Tinal and Palenewen, 1975; Kartawinata, 1975c). Large-scale timber extraction in Indonesia started only six years ago; some studies still in progress focus on the influence of logging on the biology of some primate species. During a short survey, Wilson and Wilson (1975) reported that several primate species seemed unaffected a year after selective logging. Other species might be more seriously affected, specifically the orang utan, but no direct evidence was collected. A similar conclusion was given by Chivers (1972) for the hylobatids, but he did not include details on the extent of logging and other ecological parameters. The long-term effects of logging may be detrimental for specialist primates such as the orang utan, the hylobatids and the leaf-monkeys. For instance Rijksen and MacKinnon (in preparation) found a strikingly low number of hylobatids and orang utans in a selectively logged area (Sikundur, Sumatra), probably caused by the extensive alteration of the forest structure. Moreover, it was observed that large strangling figs, which constitute one of the staple foods for these apes, are killed during such logging operations. Several primates, however, may favor certain habitats where a limited amount of disturbance inhibits the establishment of a climax plant community (Rijksen and Rijksen-Graatsma, 1975). Commercial logging has an effect which goes far beyond this and actually degrades the area for a very long period. We also found that several terrestrial mammal species were seriously affected by the logging activity (Kawamura et al., 1976b). Such is also true for several primate species, especially for the orang utan, because disturbance can lead to serious disruption of the social organization with concomittant deleterious effects for the reproduction of that species.

In general, it can be said that many problems arise as a consequence of the undisciplined actions induced by economic motives of the concessionaires. An equally large problem arises from the fact that logging operations can be permitted in several reserved areas, i.e. in the 'suaka margasatwa' or 'game reserve' (Hardjosentono, 1975). Many of these are very important reserves for orang utans, hylobatids, rhino, tigers, etc., for instance Kluet, Gunung Leuser, Sikundur, Kutai, Meru Betiri, etc. This may be one of the reasons that logging operates in some reserved areas in Indonesia (for instance Kutai reserve, as reported by Kartawinata, 1975a). Yet another problem affecting the wild stock of primates and other animals is the capture of animals for pets and the animal trade.

ADAT LAW AND SLASH-AND-BURN AGRICULTURE

The practice of Adat law exerts a great influence on the perpetuity of many reserves. Conservationists should consider using the Adat law with its many restricting regulations for conservation aims. Adat law is a tradition resulting from older tribal society in which religion regulated and controlled all aspects of life, including forest utilization. Each tribal society had its own Adat based on different social and ecological parameters. Thus, there are dozens or maybe even hundreds of different local regulations according to Adat law in Indonesia. Indeed, national law is valid and effective for all Indonesians, but in many cases groups of people still keep to their own Adat law and follow it in their everyday life. In Java this Adat law is not so prominently expressed. Activities such as hunting, harvesting forest resources or even shifting cultivation are considered activities that infringe law at the national level, but people who perform these activities often claim it as their right, according to Adat law. These conservation problems occurred in Sumatra (Brotoisworo, 1974), and reputedly throughout Indonesia. Moreover, many people use laws when they gain some benefit from it. People may hide behind an Adat law to do whatever they like, even when this law no longer regulates their daily life. Even local and illiterate people are not so ignorant as they would have one believe. Thus this is one of the major problems encountered in the conservation of forest in Indonesia; the conflict with utilization rights.

Tribal societies have been living in restricted regions for centuries and have practiced Adat law related to their particular set of environmental factors. Throughout most of that time they lived harmoniously within the environment, because they harvested resources only for sustaining the daily life of a limited number of individuals composing the tribe; the small size of the population was of paramount importance. Nowadays the equilibrium has been upset by the following factors:

1. The growing population causes an increasing demand for food and a subsequent increasing need for agricultural land.
2. Modern culture leads to a more technologically-oriented living standard.
3. Large-scale logging is practiced without consideration of rain-forest ecology and ignores long-term effects. In addition, local people often follow

the logging operations and continue to destroy the remains of the logged forest with slash-and-burn agriculture; all these factors result in a rapid acceleration of environmental depletion.

'Slash-and-burn' agriculture is a common problem in tropical countries and it is considered to be one of the major destructive agents of rain-forest habitat in Southeast Asia. This traditional method of agriculture often results in total forest destruction, turning forest land into lalang fields. In many places in Sumatra people grow perennial plants like kemiri (*Aleurites moluccana*) after they have used the land for 4-6 years, so that at least one of the functions of the forest is restored. Wirakusumah (1976) estimated that about 500-600 km^2 of lalang is made each year in East Kalimantan. These local agricultural areas are productive for approximately two years after which they are abandoned, exposed to the destructive influence of the sun and the torrential rains. Thus, on East Kalimantan alone the eroded and derelict areas will expand by 250-300 km^2 per year.

POPULATION IMPACT

One of the major problems for reserves is the use of the forest by people. The practice may be allowed by Adat law and is often forced upon them by the condition of their economic standard. The growing population and the limited sources of subsistence force people to go into the forest, where they can find various products which can be harvested easily; this is the most common problem in Java. It is aggravated by the location of reserves close to villages and surrounded by scattered human habitations; one example is the Meru Betiri reserve in East Java. This reserve is surrounded by seven villages and some coffee and rubber estates. There is little productive land, rice fields and lalang, for the number of people it must support. Sudiarto et al., (1966) estimate that in some crowded places in Java only 367 grams of rice can be provided for every person per year on less than a half hectare of productive land. Based on Sudiarto's finding, we estimated that each inhabitant near the Meru Betiri reserve can only be supported on 0.07 hectare productive land that produces only 51.38 grams rice for every person per year (Ruhiyat et al., 1976). This clearly shows that the carrying capacity of the area is surpassed, which results in a very low standard of living for the people there. The plantation estates surrounding that reserve do not significantly increase the people's income. The deprivation forces people to enter the reserve, inflicting considerable damage on the forest habitat and greatly disturbing animal life.

Obviously, it is necessary to have a broader perspective. The forest can be viewed as a subsystem of the village ecosystem. The ecosystem can then be divided into four subsystems: village, rice-field or lalang, river and forest. There is close interrelation between these subsystems and they influence each other strongly (Sumarwoto, 1974); thus the four subsystems must be considered together. The population impact cannot be solved by strict guarding or by prohibiting people from entering the forest. If this is done, the problem, a sociological one, is only shifted. The problem must be solved on a national basis through education, birth control, transmigration, etc. One of the technical possibilities is to increase the carrying capacity of the village ecosystem, while instituting efficient birth control.

PRIMATE CONSERVATION

Indonesia is very rich in primate species. There are 30 primate species, excluding *Tupaia*, distributed throughout Sumatra, Kalimantan, Sulawesi, Java and many more small islands. Thirteen species have been protected by law, some since 1931; they are *Tarsius bancanus*, *T. spectrum*, *Nycticebus coucang*, *Simias concolor*, *Nasalis larvatus*, *Cynopithecus niger**, six species of the Hylobatidae and *Pongo pygmaeus*. The other primate species have not yet been protected by law except in the existing reserves.

The legal conservation of protected animal species has improved although law enforcement is still weak. People's appreciation of the conservation of particular species supports the survival of those species. As a result of growing public awareness, improved propaganda, and legislative action of the national Nature Conservation and Wildlife Management service, several rare animals which were kept as pets were returned to the wild, for instance, orang utans were returned to the wild at rehabilitation stations in Sumatra and Kalimantan (Rijksen, 1974; Galdikas Brindamour, 1975). The aim of these programs is to diminish the capture and local trade in this highly-endangered ape. They also serve an important role in educating people and propagating nature conservation. Throughout Indonesia, hunting endangered or rare animals has decreased proportionally even though in some locations the hunting pressure on certain animals remains high. For instance, Brotoisworo and Ruhiyat (1974) reported that hunting pressure on *Presbytis cristata* is rather high in the West Bali reserve. The local people use some parts of this species for the manufacture of a traditional medicine. On the other hand, *Macaca fascicularis* is considered a sacred animal on Bali and not hunted at all. Dozens of long-tailed macaques live around temples and in the surrounding forest; for instance in Sangeh and Pulaki temples, the macaques are numerous and add to the attractiveness of the site for tourists.

In Java *M. fascicularis* and *P. cristata* are the most abundant primates; the former often raids plantations and is considered a pest, while *P. cristata* is captured in the teak forests by local people and sold as pets. *Nycticebus coucang*, a protected primate species, is also captured and sold by villagers, and is also used to make traditional medicine. *Hylobates moloch* (silvery gibbon) and *P. aygula* (surili) are restricted to West Java and there is reason for great concern that these are rapidly becoming seriously endangered. In the southern part of West Java, where gibbons can still be found, the forest has been broken up into isolated patches in many places. Habitat destruction still proceeds (Sumarwoto et al., 1976).

There are some reserves where the Javan gibbon is still found in some numbers, e.g. in Ujung Kulon and Sancang reserves. Angst (cited by Chivers, 1977) mentioned that the density of Javan gibbons in Ujung Kulon reserve is 2.5/sq km^2 and that they occur in only 20 km^2 of the reserve. In other places population densities are unknown. Some new reserves are proposed in West Java. Some of them are gibbon habitat and others may harbour both surili and gibbons.

In Sumatra, *M. fascicularis* and *M. nemestrina* are to be found abundantly in some places. In some agricultural areas they are considered pests and are hunted or trapped. These two species are also captured commercially and

---

*Sanderson (1957) classified all the Sulawesi monkeys into the species *Cynopithecus niger*. Buttihover (1917, cited by Napier and Napier, 1967), placed all the Sulawesi monkeys in the genus *Cynopithecus* which were eventually divided into two different groups of 'Macaques' (5 species) and 'Crested Macaques' (3 species); Fooden (1969) placed all the Sulawesi monkeys in the genus *Macaca* as 7 species (see also Khajuria, Vol. 3, ed.).

exported for medical research. Sumatra harbours several other primate species, most noteworthy of which is the orang utan and the hylobatids, the white-handed gibbon, the black-handed gibbon and the siamang. Even though these are protected by law, they are occasionally caught to supply the local trade and to be smuggled abroad. Smuggling was facilitated by the increase in cargo trade and timber export. With the establishment of several rehabilitation stations since 1971, this practice has decreased and many of the apes found in the towns and cities nowadays have been caught inadvertently through the effects of timber-felling and slash-and-burn agriculture rather than through deliberate hunting activities.

Problems of primate conservation on the Mentawai Islands are rather different. On these islands, the local people have hunted monkeys for food for thousands of years. They use simple hunting techniques such as bows, arrows and traps in catching monkeys, terrestrial mammals, birds and fishes. Conservation actions to alter this practice, initiated by the Nature Conservation service, have not yet succeeded. Adat law in this case exercises strict control on hunting and other activities concerned with the harvesting of forest resources. It has kept people and the environment in equilibrium for centuries. People in Siberut Island subsist traditionally on sago, taro, banana and some other vegetables. The protein supply is obtained by hunting and fishing. Sago can be obtained easily; this kind of palm grows wild in great number in the forest especially in the wet places and now they are also planted in lalangs next to taro and banana. Therefore, they never needed to clear the forest extensively. Recently through the influence of modern culture, people have started to grow some non-indigenous species. Thus in the Siberut forest there are now clearings being used for growing cash crops such as clover. The Adat hunting restrictions are more and more disregarded. Young people now hunt bilou (*Hylobates klossii*) a species that formerly has never been hunted because it was considered 'taboo'. Ritual ceremonies that preceded hunting are rarely performed nowadays.

The rain-forests of the southern islands of the Mentawai group have been allocated to timber concessions and are already being logged. Recently logging also became a serious threat on Siberut Island; it has been divided into five logging portions covering the entire region and five timber companies have been granted concessions. The Teitei Batti reserve, proposed by Tilson, covering 65 km^2 is still in the process of legislation; it is hoped eventually to extend it to 100 km^2. In addition, a considerable enlargement will be proposed eastwards to the coast (the Teluk Sarabua area, Kawamura et al., 1976a). In addition to conserving the primate species, this reserve is set aside to protect the vegetation of lowland forest, the mangrove as well as the Sarabua bay, with its extensive coral and beautiful fishes. The extent of rain-forest in this proposed reserve covers another 100 km^2.

REFERENCES

Basjarudin, H. (1976). Paper presented in the Workshop on Nature Protection and Conservation, Bogor 1976, Indonesian Institute of Science.
Brotoisworo, E. (1974). A report made for the Netherlands. Gunung Leuser Committee, The Netherlands.
Brotoisworo, E. and Ruhiyat, Y. (1974). A report made for the Council of the Netherlands Foundation for International Protection.
Burgess, P.F. (1971). *Malay. Nat. J.*, 24, 231-237.
Chivers, D.J. (1972). *In* "Gibbon and Siamang" (D.M. Rumbaugh, ed.), pp. 103-131,

Karger, Basel.
Chivers, D.J. (1977). *In* "Conservation of Non-human Primates" (Prince Rainier and G.H. Bourne, eds), Academic Press, New York.
Daryadi, L. (1976). A paper presented in the Workshop on Nature Protection and Conservation, Bogor 1976, Indonesian Institute of Science.
Fooden, J. (1969). *Bibl. primatol.*, 10, 1-148.
Galdikas Brindamour, B. (1975). *National Geographic*, 184(4), 444-473.
Hamzah, Z. (1975). A paper presented in the Symposium on long-term effects of logging in Southeast Asia, Bogor 1975.
Hardjosentono, P. (1975). A paper presented in the Symposium on the long-term effects of logging in Southeast Asia, Bogor 1975.
Kartawinata, K. (1975a). *Bio. Indonesia*, 1, 9-15.
Kartawinata, K. (1975b). *Bio. Indonesia*, 1, 17-23.
Kartawinata, K. (1975c). A paper presented in the Symposium on the long-term effects of logging in Southeast Asia, Bogor 1975.
Kawamura, S., Watanabe, K., Bakar, A., Ruhiyat, Y. and Brotoisworo, E. (1976a). Some considerations on conservation of nature in Siberut Island (unpublished manuscript).
Kawamura, S., Watanabe, K., Bakar, A., Ruhiyat, Y. and Brotoisworo, E. (1976b). A pilot study of Mentawaian primates (unpublished manuscript).
Meyer, W. (1973). *Bioscience*, 23(9), 528-533.
Meyer, W. (1975). "Indonesian Forest and Land Use Planning". University Book Store, Lex., Kentucky 40506.
Napier, J.R. and Napier, P.H. (1957). "A Handbook of Living Primates". Academic Press, London.
Pranowo. (1975). "Daftar preincian binatang-binatang yang dilidungi dengan undang-undang di Indonesia". Museum Zoologicum Bogoriense, Bogor.
Rijksen, H.D. (1974). *Biol. Cons.*, 6(1), 20-25.
Rijksen, H.D. (1975). *In* "Contemporary Primatology" (S. Kondo, M. Kawai and A. Ehara, eds), pp. 373-379, Karger, Basel.
Rijksen, H.D. and Rijksen-Graatsma, A.G. (1975). *Oryx*, 13(1), 63-73.
Ruhiyat, Y., Brotoisworo, E., Gurmaya, K.J. and Somantri, A. (1976). Laporan survey suaka margasatwa Meru Betiri, Jawa Timur, Indonesia. Lembaga Ekologi, University Padjadjaran, Bandung, Indonesia.
Sanderson, I.T. (1957). "The Monkey Kingdom; an Introduction to the Primates". Hamish Hamilton, London.
Soedjarwo. (1975). Keynote address presented in the Symposium on the long-term effects of logging in Southeast Asia, Bogor 1975.
Sudiarto, R., Sudarma, M.H., Suryono, R. and Masman Bekti. (1966). "Control program on the Citrarum-river basin and the Jatiluhur project particularly with respect to reforestation". Mimeographed, Institute of Forestry Research, Bogor.
Sumarwoto, O. (1974). Paper presented in the Symposium on Private Investment and International transactions in Asian and South Pacific countries, Sydney 1974.
Sumarwoto, O., Ambar, S. and Khan, M.H. (1976). Dense forest areas in the Citarum river basin. *Publ. Ecol. and Devel.*, 4, May 1976. Institute of Ecology, Padjadjaran University, Bandung.
Tinal, U. and Palenewen, J.L. (1975). A paper presented in the Symposium on the long-term effects of logging in Southeast Asia, Bogor 1975.
Wilson, C.C. and Wilson, W.L. (1973). Final report: Census of Sumatran primates. Regional Primate Research Centre, University of Washington, Seattle.
Wilson, C.C. and Wilson, W.L. (in press). Behavioral and morphological variation

among primate populations in Sumatra. *Yearbook of Phys. Anthrop.*
Wilson, C.C. and Wilson, W.L. (1975). *Folia primatol.*, 23, 245-274.
Wirakusumah, S. (1976). Suata tinjauan pembinaan sumber alam hayati Kaltim, usaha-usaha pengawetan dan gagasannya.

APPENDIX

List of Primate Species in Indonesia
(Chivers, 1977; Fooden, 1969; Napier and Napier, 1967;
Wilson and Wilson, 1973, 1975, in press).

No.	Family	Genus	Species
1	Lorisidae	*Nycticebus*	*Nycticebus coucang*
2	Tarsiidae	*Tarsius*	*Tarsius bancanu*
			Tarsius spectrum
3	Cercopithecidae	*Presbytis*	*Presbytis aygula*
			Presbytis cristata
			Presbytis frontata
			Presbytis femoralis
			Presbytis melaophos
			Presbytis potenziani
			Presbytis rubicunda
			Presbytis thomasi
		Nasalis	*Nasalis larvatus*
		Simias	*Simias concolor*
		Macaca	*Macaca fascicularis*
			Macaca nemestrina
			Macaca pagensis
			Macaca brunescens
			Macaca hecki
			Macaca maura
			Macaca nigra
			Macaca ochreata
			Macaca tonkeana
4	Hylobatidae	*Hylobates*	*Hylobates agilis*
			Hylobates klossii
			Hylobates lar
			Hylobates moloch
			Hylobates muelleri
			Hylobates syndactylus
5	Pongidae	*Pongo*	*Pongo pygmaeus*

MAN'S IMPACT ON THE PRIMATES OF PENINSULA MALAYSIA

MOHD. KHAN BIN MOMIN KHAN

Office of the Chief Game Warden, Kuala Lumpur, Malaysia.

INTRODUCTION

Siamang (*Hylobates (Symphalangus) syndactylus*) and gibbons (*Hylobates lar, H. agilis*) are listed as totally protected animals under the Protection of Wildlife Act, Act 76 of 1972. This is the highest form of protection for wildlife provided by legislation in the country. It puts the species in a category of animals which may not be shot, killed or taken, exported or traded. By virtue of section 51 the minister responsible for wildlife may issue permits not exceeding one to each person to collect totally protected species for scientific research. Such permits may also be issued to museums and zoos.

The category of protection for siamang and gibbons indicates clearly the intention of the Government of Malaysia to conserve her wildlife heritage. Siamang and gibbons are known to reach a number of European countries in considerable numbers. This may be interpreted in a number of ways. It is possible, in the absence of conclusive evidence, that a few of these animals may have their origin in Peninsula Malaysia, the only logical explanation being that the animals were smuggled out of the country.

Siamang and gibbons are also found in a number of neighbouring countries. It is virtually impossible to curb smuggling efficiently because a few of these countries have long coast lines. Peninsula Malaysia certainly has this problem along with East Malaysia, Indonesia and Thailand.

Endangered species convention

A number of conventions exist for endangered species, which are obviously well conceived and have a very important role to play in the protection of wildlife, providing the answer to the problem of illegal trade, smuggling and illegal exportation of endangered species. The efforts of IUCN to secure ratification by the relevant countries is a step in the right direction. More important, however, are the countries' abilities to enforce the convention: an agency is needed for the purpose. It is important for every country ratifying the convention to have a department of wildlife. It would be less effective for any country to depend on her customs or police departments as such departments already have heavy responsibilities.

The United States of America has her own convention which prohibits the importation of certain endangered species; gibbons are now listed therein.

Causes of mortality

The long-tailed macaques (*Macaca fascicularis*) are trapped in large numbers for export; the average annual export is in the region of 20,000 animals. Fewer short-tailed macaques (*Macaca nemestrina*) are exported annually. The export of leaf monkeys (*Presbytis obscura, P. cristata, P. melalophos*) is negligible.

Locally primates are kept as pets. Short-tailed macaques are widely used in the east coast states of Peninsula Malaysia for plucking coconuts. Young animals fetch good prices. In the case of long-tailed and short-tailed macaques it is possible to trap young animals without causing any harm to their parents. A young gibbon is obtained by killing its mother. For every young gibbon in captivity it is safe to assume that its mother was killed. Leaf monkeys are seldom kept as pets.

A number of macaques are killed for food. Some primates are considered a delicacy by aborigines, who possess great skill in hunting with blow pipes; they are exempted from certain provisions of the law.

The demand for some species of primate is very great; prices for them are extremely high. A gibbon can cost up to 300 US dollars. Because of the high commercial value of these animals local trappers are prepared to commit offences to obtain them for sale.

HABITAT AND POPULATION SIZE

The primates of Peninsula Malaysia utilize both primary and secondary forests. Siamang and gibbons are found in small family groups of 2 to 5 individuals. Leaf monkeys and macaques are found in primary forests but they appear to be more established in secondary forests, particularly those on the fringe of cultivated areas.

Chivers (1974), in a survey, found that siamang (*Hylobates syndactylus*) was restricted to one-fifth of the remaining Malayan forest. They may be found in the forests of the central mainland from Perak and Kelantan to southern Pahang and Negri Sembilan (Chivers, 1974; Medway, 1969; Khan, 1970). The white-handed gibbon (*Hylobates lar*) is confined to the mainland where it may be found inhabiting both primary and secondary forests at all altitudes. It is also found on the coastal areas where suitable habitat exists. Fewer records have been made of this species on the east bank of the Perak River in the north and in south Kelantan on the east (Chivers, 1974; Medway, 1969). The black-handed gibbon (*Hylobates agilis*) is found to inhabit the forests between the Mudah and Perak Rivers at all altitudes; the species was recorded in the north-west of Kelantan by the State Game Department (Chivers, 1974; Medway, 1969; Aputharajah, personal communication).

The silver leaf monkey (*Presbytis cristata*) is restricted to the west coast from Province Wellesley to Klang in Selangor. It is largely confined to the mangrove and subcoastal forest and plantations (Medway, 1969). In the State of Perak the species has been recorded as far inland as Changkat where it has become a pest to rubber. The dusky leaf monkey (*Presbytis obscura*) and the banded leaf monkey (*Presbytis melalophos*) are widespread inland in all types of forests (Medway, 1969). Both species cause considerable damage to rubber but the problem is not serious. Leaf monkeys may be found in family groups of six to twenty or more animals.

The long-tailed macaque (*Macaca fascicularis*) is found throughout the mainland. It inhabits the coastal areas, including beaches, extending its range inland to all types of forests at all altitudes. It is a common visitor to plantations and can be seen on the outskirts of most towns. It is common in most botanic gardens. The species may be found in groups varying from six to forty or more animals. The short-tailed macaque (*Macaca nemestrina*) is not as common as the long-tailed or crab-eating macaque. Nevertheless, it is widespread on the mainland ranging from the subcoastal areas to the slopes of the main range (Medway, 1969). The species is known to raid corn fields and other plantations. The short-tailed macaque is usually seen in small family groups.

Chivers (1974) estimated the population of siamang at 30,000 ± 5,000. His estimate was based on a density of 0.22 groups/km^2 with a mean group size of 3.8 animals. The population estimate for white-handed gibbons was 230,000 ± 40,000 based on a density of 0.39 groups/km^2 with a mean group size of 4.0 animals. The population estimate for dark-handed gibbons was estimated at 25,000 ± 4,000 animals based on a density of 0.39 groups/km^2 with a mean group size of 4.0 animals. The total area of forest is calculated to be 93,382 km^2.

Southwick and Cadigan (1972) reported on the abundance of non-human primates in primary and secondary forests of Peninsula Malaysia. An assessment was made of the group densities (animal/km^2) of each species except the dark-handed gibbon. Other source material include Bernstein (1968), MacKinnon (1973) and Fleagle (personal communication). The total area of forest still remaining in 1958 was 84% or 110,304 km^2.

Species	Density of Species in Each Type of Forest		Total Population
	Secondary	Primary	
Macaca fascicularis	1.54	0.37	415,000
Macaca nemestrina	0.13	-	80,000
Presbytis cristata	0.26	-	6,000
Presbytis melalophos	2.95	2.22	962,000
Presbytis obscura	0.64	0.74	305,000
Hylobates lar	0.89	1.11	144,000
Hylobates syndactylus	0.51	1.11	111,000

Estimate does not include *Macaca nemestrina* and *Presbytis cristata* in primary forest.

The primate population estimates in 1958 reflected the large tracts of forest still remaining in Peninsular Malaysia. It is important to remember that after independence in 1957 the pace of development was greatly increased. In 1958 it was reported that 74% of the land was still under forest (Wyatt-Smith, 1958). In the Second Malaysia Plan large tracts of forest were cleared for agriculture and rural development. By 1971, Aiken and Moss (1975) reported that only 91,051 km^2 of a total land area of 131,518 km^2 or about 61% were still under forest. Some 64,343 km^2 are recognized as suitable for agriculture in the Second Malaysia Plan (1971-1975). Some of this area, comprising about 28,732 km^2, is already under cultivation with the other 39,658 km^2 with agricultural potential probably scheduled for clearance (Aiken and Moss, 1975). This means that only 51.1% of the total land area will still be under forest.

Based on the same densities provided by Southwick and Cadigan, and on group sizes from MacKinnon, Bernstein and Fleagle, estimates of population of the various species, excluding black-handed gibbons, for 1975 are:

Species	Density of Groups in Each Type of Forest		Total Population (in terms of carrying capacity)
	Secondary	Primary	
Macaca fascicularis	1.54	0.37	318,000
Macaca nemestrina	0.13	-	45,000
Presbytis cristata	0.26	-	4,000
Presbytis melalophos	2.95	2.22	554,000
Presbytis obscura	0.64	0.74	155,000
Hylobates lar	0.89	1.11	71,000
Hylobates syndactylus	0.51	1.11	48,000

Estimate does not include *Macaca nemestrina* and *Presbytis cristata* in primary forest.

The losses in population of each species between 1958 and 1975 through land clearing are as follows:

Species	Population in 1958	Population in 1975	Losses in Population	% Loss
Macaca fascicularis	415,000	318,000	97,000	23.37
Macaca nemestrina	80,000	45,000	35,000	43.75
Presbytis cristata	6,000	4,000	2,000	33.33
Presbytis melalophos	962,000	554,000	408,000	42.41
Presbytis obscura	305,000	155,000	150,000	49.18
Hylobates lar	144,000	71,000	73,000	50.09
Hylobates syndactylus	111,000	48,000	63,000	56.75

The Third Malaysia Plan (1976-1980) has just begun. The Federal Land Development Authority (FLDA) has a target of 1,416 km^2 to be cleared for oil palm, rubber and other crops. Another Federal Agency RISDA (Rubber Industries Smallholders Development Authority) will clear a total of 405 km^2 in the same period for its replanting schemes. The Federal Land Consolidation and Rehabilitation Authority (FELCRA) has a target of 202 km^2 for its projects in the Third Malaysia Plan. The various state governments in Peninsular Malaysia have between them development projects requiring 809 km^2 of suitable land. It is clear that a total of 2,833 km^2 will be cleared by the end of the Third Malaysia Plan.

A number of dams are planned for flood control and to provide power for the country. Two of these are major ones that will inundate considerable area of forest when completed. The Temenggor dam has just been completed and it will soon inundate a total of 162 km^2. The proposed Tembling dam, if constructed, will inundate a total of 326 km^2 of the most important wildlife habitat of Taman Negara.

CONSERVATION ACTIVITIES

Research

Studies of primates are being actively pursued. Siamang and gibbons have received the most attention. The dusky and banded leaf monkeys were studied in the past and currently a study is being conducted on the long-tailed macaque.

Scientists have shown keen interest in studies of food habits, movements, social behaviour, populations, and so forth, in primates.

As a result we now have a large body of information on primates that is useful for making management plans. Primate studies are by no means complete; continued studies by scientists are evidence of the need for more research, and more information is needed for management. There is a need to know more, for example, about cover, tolerances of each species to changes brought about by man, and the limiting factors of each species.

It is necessary to integrate wildlife management plans with other land use planning. Vast clearings of forest for agricultural development have resulted in the extermination of a large number of primates and other wildlife species. Other forms of land use such as timber harvests also cause wildlife mortalities although to a lesser extent. Management plans should be directed towards minimising these losses.

Rehabilitation centres

Rehabilitation centres were established in East Malaysia and Sumatra for orang utans. The orang utan is endangered and it is more than justified to have rehabilitation centres for them, but there is a need to study in depth the extent of success of rehabilitated animals.

The real value of rehabilitation centres has not been fully explored. Most of the animals involved are raised from young. They become completely dependent for their food, shelter and other requirements on these centres. In most cases these captives are not given their normal food from the wild, which results in one of the major problems in adapting tame captives back to the wild state.

There are other problems apart from food. The captives are strangers to the new surrounding into which they are released. They do not know where to search for food and may not have any knowledge of the season when trees are flowering or bearing fruits. They are unable to fend for themselves.

Education

Education has a significant role to play in the conservation of wildlife in Peninsula Malaysia. It is of great concern that a comprehensive educational programme is not in existence. A well conceived extension programme is urgently needed.

A Wildlife Management School has been approved in the Third Malaysia Plan for the Game Department; a sum of 628,000 Malaysian dollars has been allotted for the project. This school will begin to function in early 1978, and will provide a training programme for Game Department staff. This agency urgently needs a highly-trained and disciplined staff to carry out its duties and responsibilities. There will be an extension programme for schools and the public.

REFERENCES

Aiken and Moss (1975).
Bernstein, I.S. (1968). *Folia primat.*, 8, 121-131.
Chivers, D.J. (1974). *Contrib. Primatol.* 4, 1-335.
Cadigan, F.C. and C.H. Southwick (1972). *Primates*, 13, 1-18.
Khan, M. (1970). *Malay. Nat. J.*, 24, 3-8.

Mackinnon, J.R. (1973).
Medway, Lord (1969).
Wyatt-Smith, J. (1958).

THE CONSERVATION OF NON-HUMAN PRIMATES IN INDIA

S.M. MOHNOT

*Department of Zoology, University of Jodpur,
Jodpur, India.*

INTRODUCTION

India is known for its diversity of fauna and flora, and its varied ecology. India's 8 primate genera and 18 species provide a striking example of faunal diversity; of these, 6 species are exclusively Indian in their distribution. Among the rest, 7 are also found in adjacent countries like Afghanistan, Pakistan, Nepal, Bhutan, Burma and Sri Lanka. The remaining 5 species, however, have restricted distributions.

Some of the Indian species have been captured and used extensively. The rhesus is an example; it has dominated the biomedical research scene for the last five decades as an important animal model. Most rhesus are still obtained from India. Some 125 breeding colonies in the USA produced a total of 2,200 live births in 1970 (Thorington, 1971) but this is just a fraction of the total number of primates used in the USA every year.

Because of their ecological vulnerability, several primate species have become rare and are threatened with extinction. The International Union of Conservation of Nature lists 40 endangered species of primate, which is more than 10% of the living primates. Many factors have contributed to this situation, including habitat destruction, increasing urbanisation, changing social patterns and attitudes, trapping, killing, hunting and epidemics. In India all these factors are relevant.

Because of the world-wide population shortage of primate species, serious thought has been given to the development of domestic breeding programmes. An international symposium was organised on breeding non-human primates for laboratory use during June 1971 (Beveridge, 1972). This was followed by a conference at Batelle's Seattle Research Center in 1972 to discuss primate utilization and conservation (Bermant and Lindburg, 1975). During November 1975, an International Winter School was organised at New Delhi on the use of non-human primates in biomedical research. The IUCN for its part has also made proposals to regulate trade in non-human primates (Harrisson, 1972). This paper summarises some aspects of Indian primates with special reference to their conservation and use in biomedical research.

NON-HUMAN PRIMATES IN INDIA

The Indian primate species vary in shape, size, population status, habitat requirements and use in biomedical research: (i) primates of special status; (ii) primates which are still abundant; (iii) primates with uncertain populations; (iv) and those which are threatened with extinction.

A monkey of special status: the rhesus

The occasional laboratory use of rhesus monkeys, *Macaca mulatta*, dates to the beginning of the century. By the late thirties usage increased, especially for poliomyelitis research. Since then, tens of thousands of rhesus have been shipped from India every year. The annual demand of rhesus in the western hemisphere reached a record number of about 250,000 macaques in 1938 (Carpenter, 1972). The flood of rhesus from India continued during the forties and fifties; nearly 200,000 individuals were exported annually in the late 1950s to the United States alone. This number decreased to about 50,000 a year to the entire world in the sixties. Between 1965 and 1974 Indian exports averaged 40,000 rhesus per year (Hartley, 1972); this is 18 times our domestic need. Since 1975, export has been restricted to 20,000 monkeys per annum, but the rhesus is still the most frequently used species, comprising 60% of the total (Nolan, 1975). Indian exports are to be reduced to a few thousand in the coming years. In summary, during the last four decades, the total harvest of rhesus must have exceeded 2 million monkeys - a number that cannot be imagined for any other species. This large-scale harvesting of rhesus resulted from the former abundance of the species, easy access to its natural habitats, easy trapping, negligible cost, and its proved hardiness in captivity.

Its extensive use in research has served not only the cause of science but of humanity at large. Unfortunately, all is not well now with this species. Until recently, its population management and conservation were neglected. The agreement for its use in an humane manner, and only for medical research and vaccine production, has not been fully followed. Rhesus are also frequently used in psychological research and production oriented programmes. For example, of about 20,000 rhesus imported by the United States in 1975 some 83% were killed during the year of their arrival. For this reason, Mohnot (1975) asked for an export ban of at least two years.

There is an increasing need in India for about 5,000 rhesus a year. Together with the regular export, this will affect deleteriously the population status if immediate measures are not taken. Southwick et al., (1975) have estimated that the rhesus population of Uttar Pradesh is about 500,000 monkeys, that there are about 176,000 births per year, and that about 60,000 individuals are available for harvest every year; whether this situation still exists today has yet to be verified. Field work by Lindburg (1971), Neville (1968), and Mukherjee and Mukherjee (1972) provides ecological data essential for conservation plans. However, the whole geographical range of rhesus should be surveyed to assess the overall population status and the number of individuals available for harvest. Only this will lead to rational plans for the conservation and preservation of the species.

It is equally important to start large-scale breeding programmes in India and in the countries using rhesus, thereby providing healthy and disease-free animals of known pedigree for research. In India the breeding programmes can be maintained in the natural habitat by enclosing large areas of forest. Although various medical precautions will be essential, costs should be low compared to those of breeding colonies in importing countries. Details are provided by Neurauter and Goodwin (1972), Carpenter (1972), Goosen (1972), Vickers (1972), Elliott (1972) and Smith (1975) that will form good bases and guide lines for breeding programmes, particularly in artificial situations. Monkeys from breeding colonies will be more expensive than ones received from the wild, but a continuous supply will be assured for biomedical research.

Whether or not breeding programmes are initiated, one important solution of the problem lies in self-restraint and judicious use. If we can compromise to some extent, the population of several of our primate species may revive to a safer limit. Napier (1968) has rightly pointed out that 'scientists faced with a research problem must choose the animal most suited to their requirements'. Perhaps the most serious question that he must ask himself is: 'Need primates be used at all for this particular experiment? Would not a white rat, a guinea pig, or a rabbit do just as well?' We should abandon the old concept of using primates because they are readily available. The biological characters of the species should be given more weight in the selection of a model. We must, therefore, seek suitable alternatives for the rhesus and other extensively used species. Even among primates, tupaias, lorises, marmosets and a few others can be used with ease and good results. This will lower the pressure on species used most till recently.

Abundant species

In addition to the rhesus, two species of tree shrews, a species of loris, two macaques and a langur may be considered abundant. Among tree shrews, the Madras tree-shrew, *Anathana ellioti*, and the common tree-shrew, *Tupaia glis*, are said to be very common. Little is known about the ecology and biology of the former, while some information is available on the latter. These 'doubtful primates' may serve as good animal models. Schwaier (1975) has been successful in establishing a breeding colony of *Tupaia*. The slender loris, *Loris tardigradus*, is found in tropical rain-forest, open woodland and swampy coastal forests of south India. It is frequently used in laboratories. Because of its sluggish nature, the demand for its eyes, and its supposed medicinal value in eye diseases, many slender lorises are killed every year. Nevertheless, all three species can be made available for biomedical research at very low cost. With some habitat management and with protection of the slender loris, the supply could be maintained for a long time.

The Assamese macaque, *M. assamensis*, occurs between 500-2,000 m altitude in the mountainous range of the Himalayas, from parts of Uttar Pradesh to Assam and down to Sunderbans in Bengal. Except for some information on its groups and their composition (Roonwal, 1949; Khajuria, 1966), nothing is known about its ecology and biology. The species, therefore, does not figure in biomedical research. The bonnet monkey, *M. radiata*, is fairly widely distributed in peninsular India, north to Satara and River Godavari. There are no censuses covering the whole range, although limited population figures, ecological data and behavioural details are known from the field (Simonds, 1965; Rahaman and Parthasarathy, 1967). Bonnets are now increasingly used in research laboratories, particularly for pharmaceutical work, as a good substitute for the rhesus.

The Hanuman langur, *Presbytis entellus*, is the most widespread of the Indian primates. Its habitats vary greatly, from snow-covered mountains up to *ca* 4,000 m (perhaps the highest primate habitat in the world), to thick forests, coastal regions, open scrub forests and arid regions. Till recently there have been only limited population surveys. These, however, do provide some picture of its present population status. Data from different study sites reveal variations in its population density, which prevent exact assessment of the total population. However, its very wide distribution, its ecological adaptability, and its moderate to large sized groups, indicate that its present population must be around a million monkeys - about double that of the estimated rhesus population. It is becoming popular among laboratory and biomedical workers, and it has been used for a variety of research, especially reproductive studies, for quite some time. The similarities of its reproductive organs to those of humans, its low cost, easy handling and regular availability make it a good animal model.

Species with uncertain populations

Six species, a tree-shrew, a slow loris, two macaques, a capped langur and a gibbon belongs to this category. The Nicobar tree-shrew, *T. nicobarica*, is confined only to Nicobar Island (Bay of Benga). No aspect of its ecology or biology is known. Likewise, the present population of the slow loris, *Nycticebus coucang*, is also unknown. It is restricted to Assam and is often sold in the local market, since it makes a good pet.

In India, the long-tailed macaque, *M. fascicularis*, is found only on Nicobar Island, from which no population data are available. Similarly, the pig-tailed macaque, *M. nemestrina*, now used frequently in biomedical research, exists in parts of Meghalaya and Nagaland, but no population figures are available. In both species populations are being depleted by hunting and removal of the forests.

The capped langur, *Presbytis pileatus*, is found in dry tropical deciduous forests and dense evergreen hill forests of eastern India, including Arunachal Pradesh, Nagaland, Meghalaya, Manipur and Tripura. This entirely arboreal langur is common in Garo Hills (Khajuria, 1962a), but its ecology and population dynamics have not been studied.

Our only lesser ape, the hoolock gibbon, *Hylobates hoolock*, lives in thick evergreen forests of eastern India, covering parts of Meghalaya, Arunachal Pradesh and Nagaland, where it is entirely arboreal. Studies of its ecology and population status are urgently needed.

Threatened species

The stump-tailed macaque, *M. arctoides*, is said to occur in dense forests up to *ca* 2,400 m elevation in Meghalaya and Arunachal Pradesh. Its rarity was discovered in 1965 when a survey carried out by Drs. Mireille Bertrand, R.K. Lahiri and George Schaller failed to locate a single group after spending several weeks, as reported by Southwick and Siddiqi (1970). Disappearance from its range was also reported by Bertrand (1969). The lion-tailed macaque, *M. silenus*, a truly arboreal species, is threatened with extinction. Kurup (1975) reported the existence of only 800 macaques organised in 55 groups over an area of about 2,800 km^2. Earlier, Sugiyama (1968) estimated its population at around 1,000 individuals and did warn of its possible extinction in the near future. Fortunately, its habitat has now been protected by law and all forest-felling operations in its range, the *sholas* in the Nilgiri hills, have been stopped.

Once a common species of the Nilgiri Hills, Western Ghats, the Nilgiri langur, *Presbytis johnii*, is threatened by destruction of its habitat, the *sholas*, because of its exclusively arboreal habits. It is also frequently killed by jungle tribes for food, fur and medicines. A tonic made from monkey parts and herbs and marketed as *Karium kurangu rasayanam* (Black Monkey medicine) is widely used in South India (Poiriev, 1971). Krishnan (1972) believes that its position has improved in recent years. Kurup (1975) also states that its former patchy distribution is now becoming continuous. Yet the extensive destruction of *sholas* in recent years for agricultural purposes has jeopardised its population. Both the lion-tail macaque and the Nilgiri langur are very specific in their habitat utilization. Thus their protection can be achieved only if forests within their range are not only protected but are replanted with the native species they use.

The golden langur, *P. geei*, of eastern India is found only in the Goalpara district of Assam and in parts of Bhutan along River Manas up to *ca* 2,400 m altitude; this interesting langur lives in dense tropical deciduous forests. From field studies by several workers, (Gee, 1955; Oboussier and Maydell, 1959; Wayre, 1968; Khajuria, 1962b; Mukherjee and Saha, 1974) it is evident that only a limited population exists. All of them have encountered only a few groups of 4-40 individuals. The geographical distribution and population status need further study.

On the basis of the observations of Lt. Col. H.S. Wood, Gee (1952) and the Editors, Bombay Natural History Society (1953), have reported the presence of the snub-nosed monkey, *Rhinopithecus roxellanae*, in Manipur and Assam. Gee believes that this species probably entered eastern India from Tibet. Groves (1970) is of the opinion, however, that 'the supposed occurence of this form in Assam, more likely refers to the subsequently discovered *Presbytis geei*'. This controversy can be solved only if surveys are undertaken in areas where Gee (1952) reported the presence of this monkey.

CONCLUSIONS

From the foregoing account it is amply clear that the future of several primate species in India is still bright. Timely action, wise use and scientifically planned management can certainly ameliorate the distressing situation that has developed in recent years due mainly to human predation. Uninterrupted envelopment of primate habitats through deforestation, urbanisation, and commercial exploitation, has curbed population growth of several of our arboreal species, five of which are now endangered.

Abundant species, like *T. glis*, *L. tardigradus*, *M. radiata* and *P. entellus*, should be cared for, protected and used in biomedical research to ease pressure on the rhesus. Rhesus population management is currently very important. Strict surveillance to minimise losses during commercial harvesting is essential, from trapping to handling, caging, transporting and shipping. The local people, villagers and farmers, have to be educated to show restraint in handling rhesus, particularly when the monkeys raid their field and orchards. Compensation may be necessary to prevent poisoning and killing. One way to minimise crop raiding would be to provision the monkeys at some distance from the fields with natural foods. This is similar to the rescue plan proposed by Bermant and Chandrasekhar (1971). Single or double - species reforestation should be discouraged. Killing of primate species in the wild (other than for scientific use) whether they are abundant, less abundant or endangered should not be allowed by framing suitable regulations.

Where information about the ecology and population levels of primates does not exist, immediate research programmes should be initiated to fill in the existing lacunae so that sound conservation programmes can take shape. Once the population status and geographical distribution are known, further work on population dynamics and carrying capacity of the habitat may follow.

Successful scientific breeding programmes in the country of origin will provide the necessary supply to fulfill the increasing research needs.

Where monkeys have settled in and around human dwellings, commercial

establishments, etc. they should be translocated to nearby jungles, or else they will be either trapped or killed. When monkeys and man are in close proximity, transmission of viral, bacterial and other diseases
Bonnet groups have been successfully relocated to nearby places from Bangalore City in the last 3 years. This can also be done with rhesus in several cities, towns and villages in northern India.

Primate exporting countries should require importing countries to ensure humane treatment and the use of primates only in those researches (biomedical and vaccine) for which they are imported. The IUCN can be authorised to enforce such regulations, by stopping exportation if they are ignored.

ACKNOWEDGEMENTS

I thank Professor M.L. Roonwal for his comments, and the Commonwealth Foundation and Jodpur University for financial assistance.

REFERENCES

Bermant, G. and Chandrasekhar, S. (1971). *Science*, 171, 628-629.
Bermant, G. and Lindburg, D. (1975). eds. "Primate Utilization and Conservation". John Wiley and Sons, New York and London.
Bertrand, M. (1969). *Bibl. Primatol.*, 11. S. Karger, Basel.
Beveridge, W.I.B. (1972). ed. "Breeding Primates". S. Karger, Basel.
Carpenter, C.R. (1972). *In* "Breeding Primates" (W.I.B. Beveridge, ed.), pp. 76-87. S. Karger, Basel.
Editors: Bombay Natural History Society (1953). *J. Bombay nat. Hist. Soc.*, 51(2), 492-493.
Elliott, B. (1972). *In* "Breeding Primates" (W.I.B. Beveridge, ed.), pp. 111-113. S. Karger, Basel.
Gee, E.P. (1952). *J. Bombay nat. Hist. Soc.*, 51(1), 264-265.
Gee, E.P. (1955). *J. Bombay nat. Hist. Soc.*, 53(2), 252-254.
Goosen, C. (1972). *In* "Breeding Primates" (W.I.B. Beveridge, ed.), pp. 88-91. S. Karger, Basel.
Groves, C.P. (1970). *In* "Old World Monkeys: Evolution, Systematics and Behaviour" (J.R. Napier and P.H. Napier, eds), pp. 557-587. Academic Press, New York and London.
Harrisson, B. (1972). *Primates*, 13(1), 111-114.
Hartley, E.G. (1972). *Br. Vet. J.*, 128, 481-487.
Khajuria, H. (1962a). *Rec. Indian Mus.*, 58(2), 121-122.
Khajuria, H. (1962b). *Rec. Indian Mus.*, 58(2), 123-131.
Khajuria, H. (1966). *Proc. 2nd All-India Congr. Zool.*, 2, 284.
Krishnan, M. (1972). *J. Bombay nat. Hist. Soc.*, 68(3), 503-555, 8 pls.
Kurup, G.V. (1975). *Indian Nat. Sci. Acad.*, New Delhi, 40-41.
Lindburg, D.G. (1971). *In* "Primate Behaviour" (L.A. Rosenblum, ed.), Vol. 2, pp. 1-106. Academic Press, New York.
Mohnot, S.A. (1975). *Indian Nat. Sci. Acad.*, New Delhi, 34-35.
Mukherjee, R.P. and Mukherjee, G.D. (1972). *Primates*, 13(1), 65-70.
Mukherjee, R.P. and Saha, S.S. (1974). *Primates*, 15(1), 327-340.
Napier, J.R. (1968). *Proc. United States Nat. Mus.*, 125(3662), 1-30.
Neurauter, L.J. and Goodwin, W.G. (1972). *In* "Breeding Primates" (W.I.B. Beveridge, ed.), pp. 60-75. S. Karger, Basel.
Neville, M.K. (1968). *Ecology*, 49, 110-123.
Nolan, M.A. (1975). *In* "Primate Utilization and Conservation" (G. Bermant

and D.G. Lindburg, eds), pp. 15-19. John Wiley and Sons, New York and London.
Oboussier, G. and Maydell, G.A. von. (1959). *Zeit. f. Morph. u okol. d. Tiere*, 48, 102-114.
Poirier, F.E. (1971). *Zoonooz*, 44(7), 11-16.
Rahaman, H. and Parthasarathy, M.D. (1967). *J. Bombay nat. Hist. Soc.*, 64(2), 251-255, 1 map.
Roonwal, M.L. (1949). *Trans. Nat. Inst. Sci. India*, 3(2), 67-122, 6 pls., 17 tables.
Schwaier, A. (1975). *In* "Primate Utilization and Conservation" (G. Bermant and D.G. Lindburg, eds), pp. 141-150. John Wiley and Sons, New York and London.
Simonds, P.E. (1965). *In* "Primate Behaviour. Field Studies of Monkeys and Apes" (I. DeVore, ed.), pp. 175-196. Holt, Rinehart and Winston, New York.
Smith, D.A. (1975). *In* "Primate Utilization and Conservation" (G. Bermant and D.G. Lindburg, eds), pp. 127-139. John Wiley and Sons, New York and London.
Southwick, C.H. and Siddiqi, M.F. (1970). *Proc. 11th Tech. Meet. Int. Union Cons. Nat.*, New Delhi, 1, 135-147.
Southwick, C.H., Siddiqi, M.R. and Siddiqi, M.F. (1975). *In* "Primate Utilization and Conservation" (G. Bermant and D.G. Lindburg, eds), pp. 25-35. John Wiley and Sons, New York and London.
Sugiyama, Y. (1968). *J. Bombay nat. Hist. Soc.*, 65(2), 283-292.
Thorington, R.W. (1971). *ILAR News*, 15(1).
Vickers, J.H. (1972). *In* "Breeding Primates" (G. Bermant and D.G. Lindburg, eds), pp. 105-108. S. Karger, Basel.
Wayre, P. (1968). *J. Bombay nat. Hist. Soc.*, 65(2), 473-477.

PRIMATE CONSERVATION IN GHANA

E.O.A. ASIBEY

Department of Game and Wildlife, Post Office Box M.239, Accra, Ghana.

INTRODUCTION

Booth (1956) lists sixteen forms of non-human primates which occur in Ghana and gives their habitats. He has indicated the broad habitat preferences and diets of West African primates (Booth, 1970). Booth's broad geographical presentation of primates in Ghana and in the rest of West Africa has been adopted for this paper. His work shows clearly that the country has a very important part to play in the conservation of the primates of West Africa (Table I, Figure 1). Only the gorilla, *Gorilla gorilla*, does not occur in Ghana. For instance, the two species of mona monkey, *Cercopithecus mona* and *C. campbelli*, both occur in Ghana. The true mona ranges from the Cameroons to Ghana and Campbell's mona occurs from Ghana to Sierra Leone. The diana monkey is confined to the west of the Volta River. The spot-nosed monkey, *Cercopithecus erythrogaster*, of Western Nigeria, is found in forest from Togoland westwards, and the red colobus, *Colobus badius*, occurs from mid-Ghana westwards (Booth, 1970).

Many species are largely frugivorous or depend on both fruits and leaves. For these species primate conservation depends, to a large extent, on habitat conservation. Man alters vegetation with his varied activities in ways that influence the distribution and survival of primates. The rates of disappearance of the primary forest and savannah and the rate of increase of cultivation affect the distribution and population of most primate species in every country. This paper deals specially with the effect of land use on the distribution of higher primates in Ghana. The conservation problems are similar throughout West Africa and even worse in some West African countries than in Ghana. Thus the Ghanaian situation is relevant to problems elsewhere in West Africa.

LAND USE

Until early this century, human activities in Ghana centered around shifting subsistence cultivation, small villages, fishing, hunting and snaring. Flint-guns were not abundant and pressures on both habitat and primate populations were relatively unimportant. The closed forest zone of Ghana occupied an area of 78,046 km^2 and formed an almost continuous forest. There were small areas under arable cultivation and larger areas of secondary forest resulting from shifting cultivation. Deforestation proceeded rapidly within this century and by the 1950s only about 28,489 km^2 of the original 78,046 km^2 of the zone carried mature unbroken forest (Foggie and Hinds, 1951).

TABLE I

Wildlife Conservation Areas of West Africa and the Status of Primates in them
(Date of Establishment is as given by 1975 UN List of
National Parks and Equivalent Reserves)

Country	National Park/ Equivalent Reserve	Area sq km	Date Established	Vegetation	Primates and their Status (a = abundant, c = common, r = rare)
Cameroon	Parc Nationale de la Benove	1,800	1968	Sudan savannah	baboon (a), patas monkey (a), green monkey (c), white-tailed colobus (r)
	Parc National du Bouba N'Djiddae	2,200	1968	Sudan savannah	baboon (a), patas monkey (a), green monkey (c), white-tailed colobus (r)
	Parc National de Waza	1,700	1936	Sahel savannah	baboon (c), patas monkey (c), green monkey (r)
	Kimbe River Game Reserve	50	1963	Montane grassland	baboon (a), green monkey (c), mona monkey (c), spot-nose monkey (r)
	Mbi Crater Game Reserve	1.3	1963	Montane grassland	baboon (a), green monkey (c)
Nigeria	Borgu Game Reserve	3,500	1966	Guinea savannah	green monkey (a), baboon (a), patas monkey (c)
	Yankari Game Reserve	1,820	1956	Guinea savannah	baboon (a), green monkey (c), patas monkey (c)
Benin (Dahomey)	W National Park	5,020	1954	Guinea savannah	baboon (a), patas monkey (c), green monkey (r)
	Parc National de la Boucle de la Pendjari	2,750	1961	Guinea savannah	baboon (a), patas monkey (c), green monkey (r)
Togoland	Reserve de la Keran	67	–	Guinea savannah	baboon (c), patas monkey (c), white-tailed colobus (r), green monkey (r)

Country	Reserve	Area	Year	Habitat	Species
Niger	Reserve de Malfacassa	1,900	–	Guinea savannah	baboon (c), patas monkey (c), white-tailed colobus (r), green monkey (r), mona monkey (r)
	W – National Park	3,000	1954	Guinea savannah	baboon (a), patas monkey (c), green monkey (r)
Upper Volta	W National Park	3,300	1953	Guinea savannah	baboon (a), patas monkey (c), green monkey (r)
Ghana	Mole National Park	4,662	1961	Guinea savannah	green monkey (a), patas monkey (a), baboon (a), white-tailed colobus (r)
	Digya National Park	3,124	1971	Transitional high forest and Guinea savannah	patas monkey (a), baboon (a), green monkey (a), spot-nose monkey (c), mona monkey (r), white-tailed colobus (r)
	Bui National Park	1,544	1971	Guinea savannah	baboon (a), patas monkey (a), green monkey (c), white-tailed colobus (r)
	Bia National Park	302	1974	High forest	diana monkey (a), mona monkey (c), spot-nose monkey (c), black and white colobus (c), olive colobus (c), red colobus (c), chimpanzee (r), white-crowned mangabey (r)
	Gbele Game Production Reserve	546	–	Guinea savannah	baboon (c), green monkey (c), patas monkey (c)
	Kogyae Strict Nature Reserve	280	–	Transitional high forest and Guinea savannah	baboon (c), spot-nosed monkey (r), white-tailed monkey (r), patas monkey (r), green monkey (r)

Table I continued overleaf

Country	National Park/ Equivalent Reserve	Area sq km	Date Established	Vegetation	Primates and their Status (a = abundant, c = common, r = rare)
Ghana (cont.)	Kalakpa Game Production Reserve	285	–	Guinea savannah	baboon (a), patas monkey (c), green monkey (c), white-tailed colobus (r), mona monkey (r)
	Shai Hills Game Production Reserve	52	–	Coastal savannah	baboon (a), green monkey (c), spot-nosed monkey (c)
	Bomfobiri Wildlife Sanctuary	52	–	Transitional forest/savannah	mona monkey (c), spot-nosed monkey (c), baboon (c), white-tailed colobus (r), olive colobus (r)
	Boabin-Fiema Sanctuary	11	–	Transitional forest/savannah	black-tailed colobus (a), mona monkey (a)
	Willi Falls	12	–	High forest	baboon (a), spot-nosed monkey (c), mona monkey (c), white-tailed colobus (r)
	Nini-Suhien National Park	16	–	High forest	diana monkey (a), mona monkey (c), spot-nosed monkey (c), black and white colobus (c), olive colobus (c), white crowned mangabey (c), red colobus (r), chimpanzee (r)
	Ankasa Game Production Reserve	343	–	High forest	diana monkey (a), mona monkey (c), spot-nosed monkey (c), black and white colobus (c), olive colobus (c), white crowned mangabey (c), red colobus (r), chimpanzee (r)
Ivory Coast	Parc National de la Comoe	11,500	1963	Guinea savannah	baboon (a), white-tailed colobus (c), green monkey (c)

	Parc National de Forest de Tai	3,500	1972	High Forest	red colobus (a), white-tailed colobus (a), diana monkey (a), spot-nosed monkey (a), chimpanzee (a)
	Parc National de la Marahoue	1,010	1968	Transitional high forest and Guinea savannah	chimpanzee (a)
	Parc National du Mont Peko	340	1968	High forest	baboon (a), chimpanzee (c)
	Reserve de Faune D'Asagny	300	–	Savannah with islands of forest	chimpanzee (c), white-tailed colobus (r), red colobus (r)
	Parc National du Banco	30	–	High forest	chimpanzee (r), spot-nosed monkey (r), white-tailed colobus (r)
Mali	Parc National de la Boucle	3,500	1954	Sudan savannah	chimpanzee (?)
Gambia	Abuko Nature Reserve	0.6	–	Guinea savannah	green monkey (a), red colobus (a), patas monkey (a)
Senegal	Parc National de Niokolo-koba	8,260	1954	Guinea savannah	baboon (a), green monkey (c), patas monkey (c), chimpanzee (r), red colobus (r)
	Parc National de Basse Casamanse	40	1970	Guinea savannah	red colobus (c), green monkey (c), mona monkey (c), patas monkey (c)

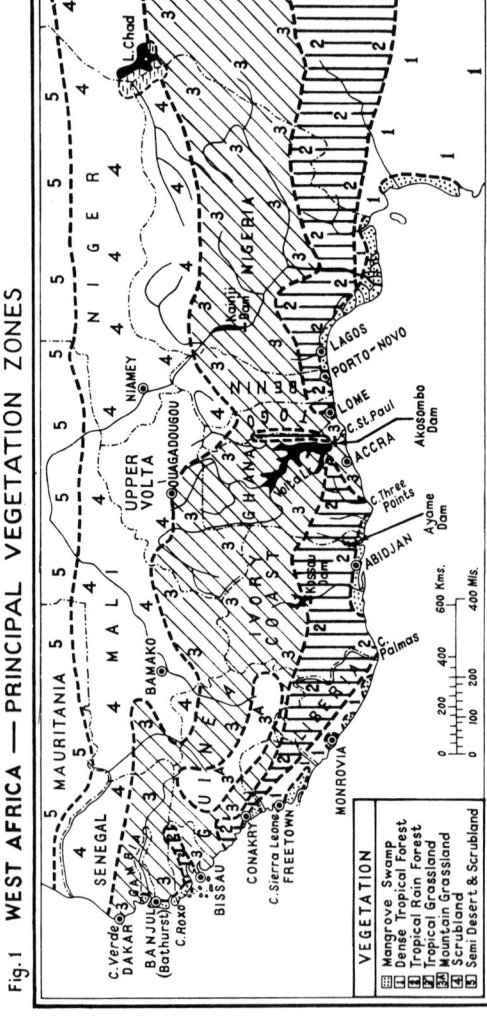

Figure 1. West Africa – Principal Vegetation Zones

Towards the end of the last century, plantation agriculture involving particularly cocoa, oil palm and rubber, became increasingly important in Ghana. Currently, plantation agriculture is a major factor in the changing scene of Ghana's land use. Over the same period of time timber industry extraction grew to become a threat to any virgin forest in the country. The export trade in Ghana timber began in 1888, nine years after the introduction of cocoa. It is now a major factor in the country's external trade. Increasingly intensive and diversified agriculture has followed the timber extraction routes in an unplanned manner. Hamlets and villages have grown into towns, and towns have become cities; foot paths have been developed into trunk roads and railways, and hitherto inaccessible territories have been opened up. These are necessary and desirable trends in the progress and advancement of the country.

Unfortunately, however, there was no recognition of the need for conservation areas. Game reserves were established in 1909 but were not effectively managed. In any event, they would have protected only the savannah monkeys. Consequently primates, which have always been a human food source wherever they abound, have been heavily hunted and exterminated locally in most parts of their range. Another stimulus for over-exploitation has been the international trade in primates (Harrisson, 1971).

It is in the light of these factors and related economic, political and social problems, that primate conservation may be viewed throughout West Africa.

PRIMATE HABITAT

Ghana can be broadly divided into forest and savannah zones. The forest zone is comprised of the tropical rain-forest proper and the deciduous high forest. The savannah zone is made up of the guinea and low savannah (Figure 2). Agricultural activities in these zones have developed at different rates. Protective forest reserves characterize the savannah zone while productive forestry characterizes the forest zone. (Productive forestry refers to the reservation of forest areas for exploitation on sustained yield basis.) These broad forest and savannah zones contain primates which are peculiar to the habitat.

Savannah zone

Water has been a limiting factor to the development of agriculture in the savannah zone. Most of the crops are largely annuals but in recent years various cash crops including rice, kenaf and cotton have been tried. Such plantations, however, are not yet a serious threat to primates. The five primates of the savannah zone (baboon, *Papio anubis choras*; patas monkey, *Erythrocebus patas patas*; green monkeys, *Cercopithecus aethiops sabaenus* and *C. a. aethiops*; and black-and-white colobus monkey, *Colobus polykomos vellerosus*) are likely to be affected by agricultural activities in two major ways. First, their habitat is destroyed for agricultural purposes. Second, animals are killed because they are agricultural pests, or as a protein source or for cash income.

The black-and-white colobus monkey is the rarest and the most sensitive of all the monkey species in the savannah zone, where it is restricted to patches of riverine forest. The green monkey also has a restricted habitat. Over the range of these species, traditional arable cultivation is giving way to commercial agriculture. This is to be expected but exacts a price particularly

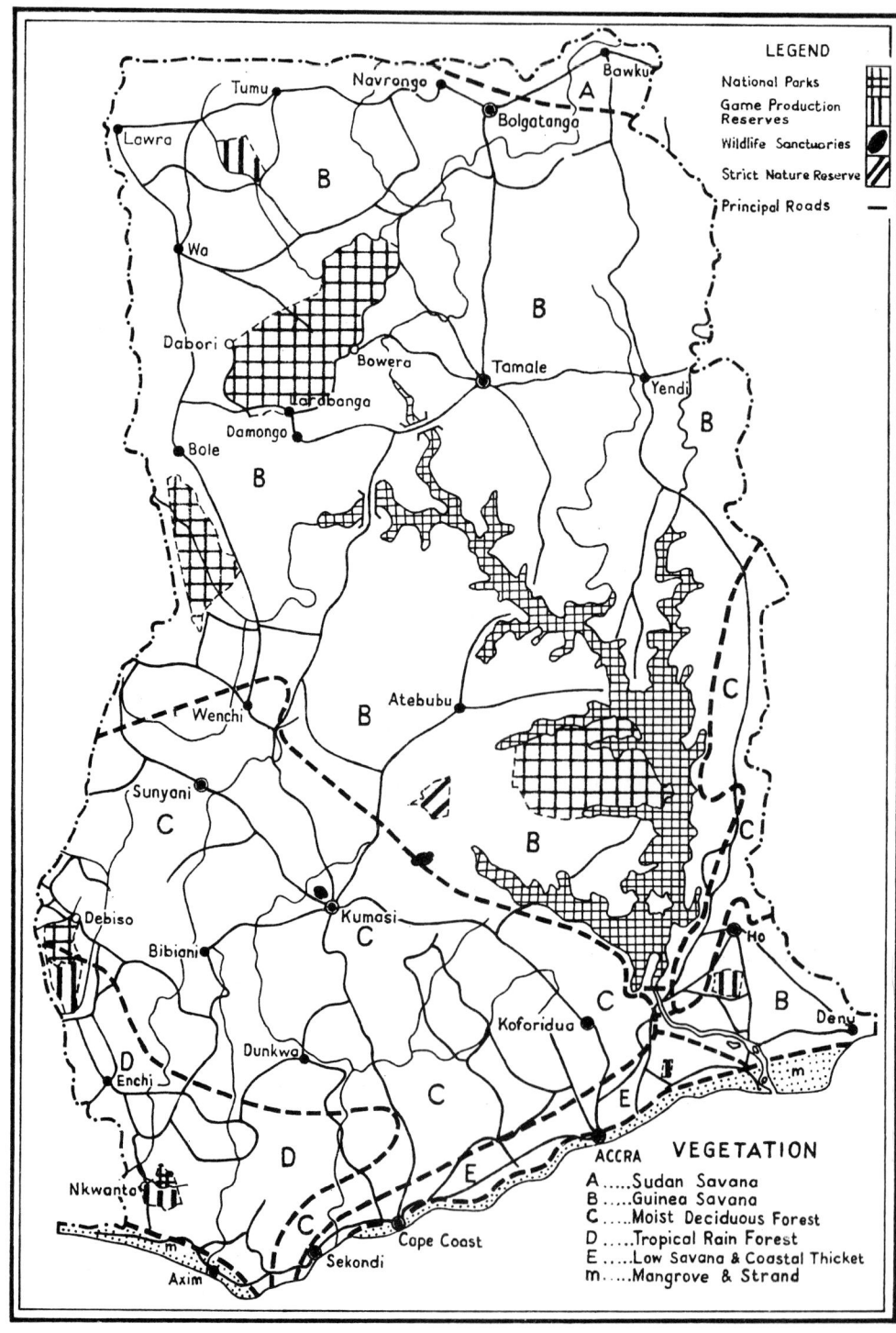

Fig. 2 GHANA — WILDLIFE CONSERVATION AREAS

Figure 2. Ghana – Wildlife Conservation Areas.

where such activities are unplanned and uncoordinated with other forms of land use. Most of the large scale modern farms in the savannah zone are being mechanised. Forests must be clear felled and stumped out to allow ploughing and the use of modern agricultural machinery. In this way most of the colobus monkey habitat in the savannah zone will be destroyed.

Even where colobus habitat survives near farm land, the species is extirpated because its meat is used domestically. The occurrence of colobus in small pockets is not compatible with exploitation on a commercial scale.

Patas monkey and baboon are the most widely distributed and least threatened by habitat destruction. Nonetheless they are agricultural pests. Patas monkey is very difficult to hunt, and its meat is almost as infrequently listed in bushmeat records as that of the green monkey. Baboon, however, appears quite frequently in bushmeat records on local markets. Farmers will control baboon numbers to the best of their ability. The ready market for its meat provides an additional incentive for control measures. It is most unlikely that poison will ever be used for pest control of any animal which is also a source of meat to humans.

The forest reserves of this zone cover major water catchment areas of the country. Thus the habitat is being preserved. Nonetheless, local people have hunting rights in forest reserves throughout the country and monkeys are hunted there. This will continue unless the rights are rescinded, usually by Wildlife Preservation Laws.

Forest zone

The remaining larger primates of the country (diana monkey, *Cercopithecus diana roloway*; mona monkey, *Cercopithecus campbelli lowei*; spot-nosed monkey, *Cercopithecus petaurista petaurista*; white-crowned mangabey, *Cercocebus atys lunulatus*; black-and-white colobus, *Colobus polykomos vellerosus*; olive colobus, *Colobus verus*; red colobus, *Colobus badius waldroni*; and chimpanzee, *Pan troglodytes verus*) occur in the high forest zone. The forest zone of Ghana and most of West Africa is under heavy pressure from agriculture. In Ghana, cocoa farming is the most important factor. Cocoa was introduced in 1879 and Ghana (Gold Coast) exported 80 lbs of cocoa in 1891. Today Ghana is the leading world exporter of this valuable cash crop, producing 29.5% of the world supply in the 1971/72 season (Anon., 1975). Small farms are individually owned. As the human population grows and the cocoa price on the world market improves, demand for more land for cocoa cultivation grows. It is common to find that city dwellers and high wage earners are absentee cocoa farmers.

The establishment of farms is generally preceded by timber exploitation to salvage valuable trees. It is often argued that intensive timber exploitation, called salvage felling, which precedes agricultural activity is a necessary evil to prevent the destruction of valuable timber by agriculture.

The intensity of exploitation has increased with the increasing popularity and diversity of species acceptable to the world timber trade. For instance, Ghana first exported timber in 1888. Before the Second World War, however, timber was imported to meet the needs of the country (Danso, 1976). Shortage of shipping space during the war forced the use of local timber whose popularity has ever since been on the increase. Before the war, only one species, *Khaya ivorensis*, was exported from Ghana (Foggie and Hinds, 1951). Now over forty species are exported (Table III). In 1947, 5,700,000 cubic feet of

TABLE II

Plants Constituting the Diet of Red Colobus Monkey at Bia National Park

Twi Name	Scientific Name	Trade Name	Economic Class	J	F	M	A	M	J	J	A	S	O	N	D	Total
Afena	*Strombosia glaucescens*	Afina	III	x	x	x	x			x	x	x	x	x	x	10
Ofram	*Terminalia superba*		III	x	x	x	x		x	x	x	x	x	x	x	11
Hyeduanini	*Guibourtia ehie*	Hyedua or Bubinga	II	x	x					x			x	x	x	6
Kokoti	*Anopyxis klaineana*	Kokoti	III	x	x	x	x		x	x	x	x	x	x	x	11
Yaya	*Amphimas pterocarpoides*	Yaya	IV	x	x									x		3
Owamma	*Ricinodendron heudelotii*	Owamma	IV	x	x	x	x									4
Akata	*Bombax buonopozense*	Akata	IV	x												1
Esa	*Celtis* spp.	Esa	III	x	x	x	x	x				x			x	7
Kyereye	*Pterygota macrocarpa*	Kyereye	IV	x	x	x	x	x		x	x	x		x	x	10
Tanduro	*Trichilia heudelotii*	Tanduro	IV	x	x	x	x			x	x			x		7
Wawa	*Triplochiton scleroxylon*	Obeches or Wawa	I	x					x	x	x	x	x	x	x	8
Okuro	*Albizia zygia*	Okuro	III	x	x	x	x			x	x			x	x	8
	Scotellia klaineana		III	x							x	x		x		4
Samantawa	*Bussea occidentalis*	Samantawa	IV	x	x	x										3
Otie	*Pycnanthus angolensis*	Otie	III	x				x	x	x	x	x	x		x	x	9
Utile	*Entandrophragma utile*	Utile	I	x	x				x	x	x	x		x	x	x	9
Onyina	*Ceiba pentandra*	Ceiba or Onyina	IV	x				x	x		x	x	x	x	x	8
Uapaka	*Uapaca guineensis*	Uapaca	IV	x					x	x			x		x	5
Dahoma	*Piptadeniastrum africanum*	Dahoma	III	x	x	x	x	x	x	x	x	x	x	x		11
	Monodora myristica	Monodora	IV					x								1

Primate Conservation in Ghana

Local name	Scientific name	Trade name	Class								Total
Wawabima	*Sterculia rhinopetala*	Wawabima	III	x		x	x	x	x	x	6
Lovoa	*Lovoa trichilioides*	African walnut	I	x x x	x x x		x x				8
Onyinakoben	*Bombax brevicuspe*	Onyinakoben	IV	x x x		x	x				5
Duabaha	*Hexalobus crispiflorus*	Duabaha	IV	x							1
	Canarium schweinfurthii		III	x							1
Koroma	*Klainedoxa gabonensis*	Koroma	IV	x x x x	x x x x x			x			10
	Baphia pubescens		IV	x x							2
Papao	*Afzelia bella*	Papao	III	x					x		2
Watapuo	*Cola gigantea*	Watapuo	IV	x							1
Funtum	*Funtumia africana*	Fruntumia	IV	x x x		x	x				5
Kusibini	*Diospyros sanza-minika*	Kusibini	III	x x							2
Akasaa	*Chrysophyllum albidum*	Akasaa	IV	x				x			2
Odum	*Chlorophora excelsa*	Iroko or Odum	I	x	x	x					3
	Tetrapleura tetraptera		IV	x							1
Sapele	*Entandrophragma cylindricum*	Sapele	I	x	x x x						4
	Trichilia megalantha		IV		x	x x					3
Emire	*Terminalia ivorensis*	Emeri or Imdigbo	I	x x	x	x					4
	Aningeria robusta		IV	x					x		2
Awiemfuosamina	*Albizia ferruginea*		III	x x x	x x x x x				x		7
Aprono			II	x	x x x x				x		5
Otanuro	*Trichilia tessmanusi*		IV	x x x x							4
Potrodom	*Erythrophleum ivorensis*	Potrodom	III	x x x x x							5
Korokon	*Scotellia chevalierii*		III	x							1
Mahogany	*Khaya anthotheca*	Anthotheca	I		x	x			x x		4
Ayan or Bonsamdua	*Distemonanthus benthamianus*	Bonsamdua	III		x x						2

Table II continued overleaf

Twi Name	Scientific Name	Trade Name	Economic Class	Period of the Year (months) J F M A M J J A S O N D												Total
—	*Trichilia martineaui*		IV					x									1
Mansonia	*Mansonia altissima*	Mansonia	II					x									1
Hatrohotro	*Hannoa klaineana*	Hatrohotro	IV									x					1
Bediwonua	*Canarium schweinfurthii*		III									x	x				2
Akumadua	*Nesogordonia papaverifera*		IV											x			1

TABLE III

Composition of Economic Tree Species in the Diet of Red Colobus Monkey at Bia National Park over a Period of 12 Calendar Months

Economic Class	No. of Tree Species Recognised in:		Red Colobus Diet		Frequency of Use	
	Ghana	Bia National Park	No.	%	No.	%
Class I	14	11	7	63.6	40	17.24
Class II	12	4	3	75.0	12	5.17
Class III	24	18	18	100.0	99	42.67
Class IV	174	127	23	18.1	81	34.91
Total	224	160	51	31.9	232	100

valuable timber were exported by Ghana (Foggie and Hinds, 1951). In 1972, 66.33 million cubic feet were exported. Also in 1972, in addition to exported timber, 351 logs were rejected and 20,342 logs were locally sawn. During that year, log congestion at the harbour was worse than ever before (Francois, 1974). There is no record of general timber wastage, but logs stranded by the roadside, and remains left in the field must amount to a considerable volume.

The increase in the importance and diversity of timber species on the world market has come at a time when timber is also required as an important source of foreign exchange for the modern development of the country. This encourages timber exploitation both within and outside forest reserves. Local wood based industries are growing and causing an increasing demand for timber. The home market too desires the famous timber species of international repute. Unfortunately most of the unreserved forest estates of the country have already been exploited and farmed. Such farmed land becomes private property. However, tree farming is not commonly practised in Ghana. Consequently the forest reserves, primarily established to meet local demand for forest products (Foggie and Hinds, 1951) are now under heavy pressure to meet the demands of a very attractive external trade. The timber felling cycle of these reserves used to be 25 years but is now 15 years (Anon., 1970). It is interesting to note that it takes about 80-120 years for the forest region to regrow valuable timber trees (Keay, 1973).

Vicious circle

With the spread of cocoa farming across the forest zone from east to west, salvage felling increased in tempo, followed by what amounted to clear felling related to farming activity. Salvage felling had to be accelerated ahead of farming. As timber extraction routes opened up the forest country, farmers followed and set up their farms. Food crops produced in the initial stages of the establishment of cocoa plantations are easily transported on timber trucks to population centres. In this way there was no place too remote to be worth cultivating. By the time the lumbering was over, the cocoa trees were bearing fruits and the cocoa economy was there to stimulate road construction and further development of the area. In this way salvage felling was pushed hard. It increased over the years until at the present day, if it were not for the forest reserves, Ghana would presumably be importing timber. This process is not peculiar to Ghana. Most of West Africa has experienced this escalation of salvage felling which has been sustained by the ever increasing world demand for tropical hard wood. A traveller through the West African Region is surprised by the lack of tropical high forest. Instead there are large tracts of thickets and secondary forest. It would take 60 years for these tracts to yield trees of comparable size to what is being exploited today (Keay, 1973). Such reestablished forest, however, would be devoid of its original primate species diversity and population.

Primate conservation in the forest zone like the savannah zone is closely related to habitat conservation. The situation in the forest zone is even more significantly related to forest conservation because most of the primates concerned depend on the flowers, fruits and leaves of the forest trees for their food.

The relationship between timber trees and primates of the high forest

Timber trees are usually classified according to their demand and popularity in the timber trade. In Ghana, the classification recognises four classes. The best known and the most popular and important species are in Class I and the least known are in Class IV. It is important to note that (Tables II and III) 11 out of 14, 4 out of 12 and 18 out of 24 economic timber trees of Classes I, II and III respectively, occur in the Bia National Park (Owusu and Baidoe, 1970). It is generally accepted that all other trees not listed in Classes I to III belong to Class IV (Danso, 1976).

According to a preliminary analysis there are 51 tree species, abundant in undisturbed forest, which are eaten by red colobus (Table II). In Table II each tree species gets one point for every month in which the red colobus was observed to be eating any part of it. Thus the maximum number of points per species is 12 following the method for ranking the diet of grasscutter (Asibey, 1974a). The trees most frequently used are in the economic classes III and IV. This however, reflects the availability of these timber species and may not define to what degree the species of Classes I and II are important. For example, a higher percentage of the available species in Classes I-III are utilised by red colobus than in Class IV, indicating the possible importance of these species (Table II). Besides, vines comprise approximately 20% of the red colobus diet (Rucks, 1976).

A study of the red colobus and the diana monkey has shown that a large part of their diet is dependent on timber species all the year round (Tables II and III). It is difficult to say how easily these species can survive on other non-timber species because they are so easily eliminated in disturbed forests by hunting. It has often been argued that lumbering by itself has no adverse effect on the animals of forest habitat. Unfortunately, there is nowhere that lumbering alone has influenced the primate populations. It should however be possible to evolve a cultural system in which lumbering is compatible with primate conservation, provided that only the diet of primates is affected by lumbering.

It is proposed to investigate the effect of lumbering *per se* on primates in the high forest zone. The study will be conducted in national parks and game reserves adjacent to the parks. The game production reserves will be logged but the animals will not be exploited. No farming or any other form of human activity which would constitute interference will be allowed. Most of the sensitive primate species appear catholic in their diet, and they may be able to thrive on uneconomic timber species and other plants. If so, forest reserves can be constituted as game production reserves, the hunting rights bought, and more land will be available for both timber production and primate conservation. Such multiple land use will be in the best interest of all concerned.

In Ghana, the legal framework is already in existence. Under Ghana laws, the mere entry into any class of game reserve without a written permit constitutes an offence. There is also an instrument for compulsory acquisition of reserves. It is hoped that current and future research work in the game production reserves and parks will enable us to understand better the importance of different plants to the primates. We would then be able to apply the the laws with greater confidence, and manage and conserve the primates and their habitat more effectively.

Under the increasing pressure for land for agriculture and other uses incompatible with primate conservation, such knowledge is vital to the conservation of West African primates in general and those of Ghana in particular. Meanwhile, since timber species constitute an important component of the diet of these monkeys (Tables II and III), it is reasonable to advocate a reduction if not a complete halt to the exploitation of sensitive primate habitats. It is recognised, however, that the world is now interested in the conservation of the tropical rain-forest in general. The success of this movement will not necessarily guarantee the conservation of primates unless their exploitation is prohibited.

THE ROLE OF CONSERVATION AREAS IN PRIMATE CONSERVATION

In Ghana, national parks and various classes of game reserves are collectively known as wildlife (including habitat) conservation areas. The wildlife conservation policy of the country, manifest in the current network of conservation areas of the country, was developed after independence. Conservation refers to the management and the wise use of wildlife resources on a sustained yield basis. It is too early to know precisely its direct impact on the economy of the country. At the onset, the government appreciated the difficulty in getting land owners voluntarily to allow the use of their land for the purpose of wildlife conservation. Consequently, it was deemed necessary that conservation areas be compulsorily acquired. The exercise of this power rests with the head of State. Compulsory acquisition makes it obligatory on the State to pay compensation to land owners. Where villages are affected they have to be resettled at the expense of the State. The creation of conservation areas in Ghana is therefore a very expensive business. Nonetheless, it has been possible to constitute national parks and game reserves in all the major vegetation zones of the country (Figure 2).

Notwithstanding Ghana's desire and determination to conserve wildlife, any major conflict between primate conservation and agricultural development will be resolved in favour of agriculture. Primate conservation in both the forest and savannah zones will depend on the establishment of conservation areas where the continued existence of primates can be accepted in the context of land use. Should this appear impossible, it will be necessary to identify areas of incompatibility and find means of ameliorating the conflict.

Most of the forest reserves in the high forest zone were either established or earmarked for establishment during the colonial era, particularly in the early 1930s. The primary purpose of the reserves was to ameliorate climatic conditions, protect water catchment, protect soil and enhance agriculture, particularly cocoa farms. In effect forest protection was to serve the major interest of agriculture. This attitude persisted over the years and pressures have become increasingly heavy to change the status of reserves in order that the land may be cultivated. Forest reserves in the savannah zone included areas for plantation forestry to provide fuel wood and building poles. Since wild animals were and still are a very important source of food to the farmers and even now to the urban dwellers, it was inevitable that the people would be allowed to retain their hunting rights in forest reserves. In these days heavy demand for meat makes well-stocked reserves attractive to hunters. If and when hunting rights in forest reserves are abolished, forest reserves which would still be rich in primate fauna will be additional areas of significance to national parks and game reserves. Where plantation forestry is practised, however, the animals are exterminated in the initial stages of the project. Such plantations will have the same adverse effects as agriculture on primate conservation.

It is of interest that throughout West Africa, the important national parks and game reserves are in the savannah zone (Happold, 1973). The savannah primates are well represented in all the parks and reserves within this geographical range (Table I). It is therefore reasonable to say that as long as the reserves and parks in the savannah zone of West Africa are effectively managed, troops will survive and will be conserved.

In the forest zone however, the situation is not so rosy. National parks and game reserves are much fewer. Yet it is in this zone that the greatest diversity of primates can be found. The primates form a major source of animal protein and hunting pressure is high even in the numerous forest reserves. With the current rate of disappearance of the high forest still increasing, the primates have been safe only where they have been protected by customary laws. Under customary laws, groves, which to all intent and purpose are wildlife sanctuaries, were established many years ago. The groves are often rich in black-and-white colobus. Spot-nosed monkey too may be present. Spot-nosed monkeys have also been able to persist in secondary forest and thickets close to heavy human population and activities. The mona monkey too has persisted in some thickets and secondary forest along with the spot-nosed monkey. The more easily they are hunted, the sooner most of the primates are wiped out to provide meat for the ever-widening bushmeat market (Asibey, 1974b).

The best-known parks in the forest zone are the Tai of the Ivory Coast and the Bia of Ghana. Ghana has recently established an additional national park and game production reserves in this zone which are equally rich in primate fauna.

Faunal surveys of the forest zone in Ghana have shown that the Bia National Park offers an important habitat for primates and that there is a good chance for the survival of the species which occur in this park. The park, which has an area of 230 sq km, is in the deciduous forest zone (Figure 2). The new park and a game production reserve of 165 sq km and 343 sq km respectively are in the tropical rain-forest proper. They contain the same primate fauna as the Bia National Park and its game production reserves.

The Nini-Suhien National Park and the Ankasa Game Production Reserve are additional strongholds for primate conservation in West Africa. The same species of primates are seen more frequently in these new conservation areas than in the Bia National Park. The frequency of observation of primates in the Bia is 0.84 animals per hour as compared with 1.16 animals per hour in the new conservation areas (Martins, 1976).

These high forest conservation areas of Ghana appear to be the richest primate conservation areas in West Africa (Table I). Granting that the Bia and the Nini-Suhien National Parks are guaranteed to remain untouched, all the high forest primates of Ghana are represented and are at the moment in such a balanced ecosystem that the unit can be regarded as unthreatened as long as the park staff work effectively to keep poachers out and as many troops as possible remain permanently in the conservation areas.

PRIMATE EXTERMINATION

In Ghana and other West African countries, primate meat is eaten by many people. The ready market for primate meat is an incentive to hunters. Many farmers too depend on this and other sources of bushmeat for most of their animal protein supply. The extermination of primates can be attributed, in no small measure, to hunting for the pot and for cash income (Asibey, 1974b). Weapons are essentially tools for primate extermination. These tools of destruction have become more efficient; shot guns and even automatic weapons have replaced the flint guns and bow and arrow. Almost every successful farmer owns a gun. In the savannah zone, the possession of a gun is a sign of manhood. Many prosperous urban dwellers own weapons which are illegally leased out to rural dwellers in return for bushmeat. In the past hunters used to cut their own hunting trails. These days most hunters need not cut trails. Timber extraction routes are common and these routes are fully utilized. Recently the Forestry Department began to set out permanent sample plots in a number of forest reserves which had not been exploited. Hunters took advantage of the cleaned transect lines and wiped out large mammals of the study areas.

All colobus monkeys and the chimpanzee are fully protected in Ghana. Nonetheless they are heavily poached outside national parks and game reserves. The red colobus monkey is the most susceptible of the forest primates to hunting pressure. It is the first species which becomes extinct in local areas where hunting is common. Being a highly vocal animal of the upper canopy, it can be heard for more than 370 metres. This enables hunters to locate troops. When disturbed, the troops are noisy and flee in single file. They do not disperse. A fast hunter can therefore move ahead of them, lie in wait and shoot them one by one. The diana monkey is also susceptible to hunting pressure. It appears in reduced numbers where red colobus has been eliminated. It and the red colobus may be absent where the other monkeys are present in reduced numbers. It is vocal and can be detected from a distance but it has the ability of effective dispersal when a troop is disturbed. The other monkeys are less vocal, can keep in lower canopy or even on the ground when in danger, and are difficult to detect by hunters. Consequently they survive longer than the red colobus and the diana monkey. The mangabey and the chimpanzee seem to be the only primates which tend to wander in and out of the parks and reserves to any significant extent (Rucks, 1976). Mangabey is the only forest primate usually guilty of raiding farms (Cansdale, 1960). Hunting mangabey and chimpanzee in the parks and game reserves is very difficult since they travel on the ground, are very alert, and are far too fast to follow. It seems that most of them are shot on farms outside the parks and reserves where their natural cover of the understory has been removed (Rucks, 1976).

It is important to know the home range of most of these primates and ensure that parks and reserves are large enough to contain viable troops which can be self-sustaining. Trends in land use have indicated that the parks and reserves will ultimately become islands in the ocean of farms. Primates that will tend to wander outside the conservation areas will probably be shot.

PROBLEMS IN ESTABLISHMENT OF NATIONAL PARKS/GAME RESERVES

Conservation of primates and all other wild animals is expected to have economic justification. In the late 1950s and early 1960s, there was concern that modern African states might consider wildlife conservation as reminiscent of primitiveness and a hindrance to modern development. The past decade has however seen an increase in the establishment of conservation areas throughout modern African states, thereby proving their willingness to accept responsibility for the conservation of their wildlife heritage (Asibey, 1963).

Conservation refers to the wise use of resources on a sustained yield basis. Unfortunately, until recently, conservationists themselves have had the tendency to muddle the concept of preservation which characterized the national park philosophy in the nineteenth century and was applied in the 1933 London Convention for African Wildlife Preservation.

Because of the changing economic scene, wildlife preservation *per se* is difficult to justify in modern Africa. It is doubtful that such wildlife preservation policies can continue anywhere, for that matter. Because of human exploitation and ever increasing mobility, romantic attitudes towards renewable natural resources will be discarded.

It is legitimate to insist on wildlife conservation in the context of national land use policy. Unfortunately such policies do not exist and anything near it is but lip service. The cardinal point is economics. There is also the question of land ownership. Various laws have been promulgated, vesting land in the State. In practice, however, land is still communally owned or vested in chiefs and heads of families. In Ghana, land ownership is no hindrance to the establishment of national parks and conservation areas. Unfortunately, in the high forest zone, all suitable primate habitats which are still rich in number and diversity of species, are already committed to other forms of land use. The wildlife conservation movement that swept the savannah zone reached the forest zone too late. The zone has many forest reserves, but the reserves have been leased to various timber companies and are earmarked to be logged. The companies have been paying royalties for the many years that they have held the concessions. Should the reserves be taken from them, their rights must be purchased. The problem is, however, not as critical now that the government holds a majority share in most of the large, companies. Other rights including fishing and hunting must be rescinded. This raises serious social, political and economic issues that affect the survival of many people who depend on such areas for their meat and other day to day supplies of essential domestic needs.

It is very difficult to reconcile all these conflicting uses. The establishment of high forest parks and game reserves like Bia, necessitates political and economic decisions at a high governmental level. At best, the professional man can show all the available scientific, cultural, aesthetic and environmental bases for such decisions but it is nearly impossible to draw up the balance sheet. Even when the government is keen to establish such parks and reserves, international financiers and experts are quick to advise against it, and this makes the situation more complicated. This unhealthy attitude has been prevalent in wildlife conservation in West Africa as a whole and Ghana in particular. It is generally believed that parks and reserves must be viewed in the context of tourism as it pertains to East and other parts of Africa, or it is economically unwise to establish them. In the majority of cases, the role of the wildlife resources in the local economy has not been studied and although qualitative knowledge is available, it is difficult to accept in the absence of statistical data (Asibey, 1974b). This bedevils the future of even the better-known national parks of Africa. The existence of these is precarious in the face of the increasing human population, demand for agricultural and grazing lands, and the cash economy commanded by these forms of land use (Myers, 1972).

When it comes to primate conservation, a different method is required to introduce tourists to the high forest as a recreational area. It is to be hoped that suprasonic man will have the patience, humility and self-discipline to sit and wait in silence to observe and enjoy the wonders of primates in their natural environment. At the moment, high forest, coupled with its high humidity and crawling insects, is not particularly attractive to most people wishing to see spectacular African wildlife. It is hoped that with time this will be equally popular. Time is not on our side, however. Ghana requires capital for the development of its essential services to ensure the survival of its people. The nation has no romantic attitude towards wildlife. We must develop the means of introducing tourists to high forest parks and primate viewing, to the point that the parks have an economic reason for existing.

Fortunately, wildlife conservation in Ghana has not been based on tourism or economic feasibility studies *per se*. The wildlife conservation movement in Ghana was a governmental desire to try and save some of its wildlife heritage. Such moral obligations cannot be expressed in economic terms. Left to most international financiers and advisers, Ghana would not have even attempted wildlife conservation in the savannah zone. Such financiers and advisers obviously did not understand the point at issue. To them it is too late to catch up with other parts of Africa. Besides there is neither the diversity nor abundance of wild animals to justify wildlife conservation economically and its cost in terms of land, other forms of capital and manpower. Many West African countries are in the same situation. Perhaps it is time for those concerned about the future of the world's wildlife to reconsider their concepts in the context of local conditions.

This problem has come up again in the quest for national parks in the forest zone where the park will freeze valuable land and invaluable timber resources. The question is not as settled in the forest zone as it is in the savannah zone. There are many wood-based industries requiring timber. There is also the need to export timber. If research proves that normal lumbering will not adversely affect the primates in such parks and reserves, then there is no cause for anxiety. If not, then there will be a conflict between conservation and logging. So far the Ghana Government has had the political and economic will to let the high forest park remain. The increasing shortage of timber land should give one cause for concern, despite the incredible backing and firm stand so far taken by the government.

These factors make it necessary to search continually for a broader base for the sustained use of parks and reserves. Perhaps the Man and Biosphere Programme will provide additional useful knowledge to strengthen the feeble knees of wildlife conservation.

As a policy, an effort is being made to link parks and reserves to the overall development of the areas in which they are located. So far this has been successful. It is hoped that the same success will attend the course of national parks and reserves in the high forest zone. It is extremely difficult, however, to break into the highly individualised and well-paid agricultural and timber operations of this zone because wildlife conservation yields poor direct financial returns and it conflicts with many personal interests and private gains. So far there has been no doubt that there is the will and the courage of the government to allow this form of land use, however unpopular it may be, on the grounds of its moral obligations to posterity. As long as this policy stands, reserves and parks will continue to be considered a form of land use related to rural development, and it can be confidently said that the primates of Ghana will continue to survive for the benefit and enjoyment of humanity and Ghana will continue to play her role in primate conservation in the world.

ACKNOWLEDGEMENTS

Mr. S.R.K. Loh, Deputy Chief Cartographer and Mr. E.H. Andrews, Chief Lithographer, both of the Ghana Survey Department, helped in compiling Figures 1 and 2 and also supervised their production.

REFERENCES

Anon. (1970). Seminar on Management of the High Forest of Ghana. Mimeographed. 26 pp. Ministry of Lands and Mineral Resources, Accra.

Anon. (1975). "Africa South of the Sahara 1974". Europa Publication, London.

Asibey, E.O.A. (1963). *Wildlife*, 3, 20-25.

Asibey, E.O.A. (1974a). Some Ecological and Economic aspects of the Grasscutter, *Thryonomys swinderianus* Temm. (Mammalia; Rodentia, Hystricomorpha) in Ghana. Unpublished Aberdeen University Ph.D. Thesis.

Asibey, E.O.A. (1974b). *Biological conservation*, 6(1), 32-39.

Booth, A.H. (1956). *W. Afr. Sc. Ass. J.*, 2(2), 122-133.

Booth, A.H. (1970). "Small Mammals of West Africa". Longman,

Cansdale, G.S. (1960). "Animals of West Africa". Longman,

Danso, C.K. (1976). Stocking and distribution of prime and secondary species in the Ghana Tropical High Forest. Mimeographed. 23pp. Forestry Department, Kumasi.

Foggie, A. and Hinds, J.H. (1951). *In* "Management and Conservation of Vegetation in Africa". Bulletin No. 41 of the Commonwealth Bureau of Pastures and Field Crops. Penglais Aberystwyth, Wales.

Francois, J. (1974). Ghana annual report of the Forestry Department Ministry of Lands and Mineral Resources for the calendar year 1972. Mimeographed. 44pp. Forestry Division, Accra.

Happold, D.C.D. (1973). "Large Mammals of West Africa". Longman,

Harrisson, B. (1971). "Conservation of Non-human Primates in 1970". Karger, Basel.

Keay, R.W.T. (1973). *African Research and Documentation*, No. 1, 5-10.

Martins, C. (1976). Report on a survey of the Ankasa River Forest Reserve. Mimeographed. 17pp. Department of Game and Wildlife, Accra.

Myers, N. (1972). *Science*, 178, 1255-1263.

Owusu and Baidoe (1970). Bia Group Forest Reserve Working Plan. Mimeographed. Ministry of Agriculture Forestry Division, Accra.

Rucks, M. (1976). Notes on the problems of Primate Conservation in Bia National Park. Mimeographed. Department of Game and Wildlife, Accra.

THE CONSERVATION OF PRIMATES IN THE UNITED REPUBLIC OF CAMEROON

JOSEPH AWUNTI

Vice Minister of Agriculture, United Republic of Cameroon.

It is a great pleasure for me to represent the United Republic of Cameroon in this Sixth Congress of the International Primatological Society. of State, His Excellency President Ahmadou Ahidjo, immediately approved of a Cameroonian delegate because of his genuine interest in the conservation of Cameroon wildlife. We are always ready to co-operate with the International Primatological Society, just as we have co-operated with all other international organisations with similar objectives.

In 1932 the first Cameroon reserves were established by the Governor of the then colonies. So far, a total of 23,000 km^2 of wildlife parks have been established. It is government policy to increase this area, and the Forestry Department is actively pursuing this policy. Thus the United Republic of Cameroon is complying with the aims of the IUCN to conserve whole areas for wildlife. However, special attention must be paid to the conservation of special species such as the primates which are threatened or endangered.

SUITABILITY OF THE UNITED REPUBLIC FOR PRIMATE CONSERVATION

Our primate fauna is among the most numerous and most diverse in the world, ranging in size from the tiny bush-baby to the western gorilla. Habitats vary from dense rain-forest and swamp in the south to arid sahel regions in the north, providing diverse ecological situations for our wildlife. Politically, the United Republic of Cameroon has a stable administration, a necessary factor for international assistance in conservation and involvement at a very high level.

The President himself, a strong believer in wildlife conservation, is extremely anxious to encourage conservation. He has promulgated a number of ordinances, decrees, orders and circulars for the conservation of our wildlife resources. He has made untiring and ceaseless efforts to dissuade the abusive use of firearms for the destruction of our wildlife.

Cameroon has a Wildlife College at Garoua for the training of all francophone middle-level cadres in wildlife. It is hoped that the school will soon train graduates in biology and enable francophone African countries to have policy-makers in wildlife conservation. The school is headed by a Cameroonian bilingual Wildlife Biologist assisted by eight Cameroonian senior staff.

In Cameroon, ecological and conservation research is undertaken by the National Office of Scientific and Technical Research (ONAREST) which has an Institute for Agricultural and Forestry Research. This Institute has a Centre of Forest Research and grants all authorisations to wildlife researchers interested in conservation.

Finally, Cameroon is officially a bilingual country with French and English being the official languages. This gives it a unique opportunity to facilitate communication between francophones and anglophones. The bilingual status of the United Republic of Cameroon facilitates and encourages research by scientists from countries overseas which speak either French or English.

PRESENT EFFORTS IN WILDLIFE CONSERVATION

(1) Edea Park

In 1971 the Government of Cameroon was asked to convert the Douala-Edea and Korup Reserves into National Parks and to create a Primate Research Centre in these Parks. The President of the United Republic, after a careful consideration of this timely proposal, ordered that these two areas should be conserved as wildlife parks for scientific purposes. This meant the immediate prohibition of wildlife exploitation.

In keeping with the Presidential Order, timber exploiters were evicted from the Douala-Edea Reserve. It was also necessary to re-educate the people who were to derive immediate benefits from the timber exploiters so that they appreciated the dire need of the total conservation of this reserve.

In 1974, the Government appointed a Conservator in charge of the Douala-Edea wildlife reserve in order to ensure the proper protection of the primates of this area.

During the period 1974/1975 the sum of about 2,000,000 francs CFA was allocated to the Conservator for initial surveys. Sites have been designated for the construction of Guard Posts at the villages of Pongo-Songo on the Sanaga River and at Elogbatindi on the Edea-Kribi road at the southern end of the reserve. Dr. Gartlan has been carrying out research activities in the Douala-Edea Reserve under government authorisation and he is expected to submit his report soon. It is our hope that his report will give us food for thought in our wildlife policy.

(2) IUCN Project 1089

The majority of the national parks in Cameroon are found in the savannah areas, in the north of the country. There are, however, a number of tropical rain-forest/game reserves in the south. In view of the paucity of tropical rain-forest national parks, conscious of the willingness of the Government of the United Republic of Cameroon to transform Douala-Edea and Korup Reserves into National Parks, in keeping with the opinion, 'that the conservation of tropical rain-forest areas should be the top priority of the International Primatological Society' and in recognition of the need to protect the great apes, which exist in these reserves, the IUCN submitted Project Number 1089 to the Government early in 1975 for study.

This project will cost IUCN about 16,000,000 francs CFA and the Cameroon Government 11,800,000 francs CFA. This project was sent to the appropriate technical services for study. The Government is now studying the proposals of the technical services and it is hoped that as soon as possible the project will commence. The project, I presume, will be headed by Dr. J.S. Gartlan and it is expected that its results will provide useful information for the management of the game reserves in question.

(3) Gorilla conservation

Cameroon is among the few nations in Africa which are fortunate to have the lowland or western gorilla (*Gorilla gorilla gorilla*). We are conscious of the endangered state of this very precious rare species. Consequently, every effort is being made to conserve it. Thus the gorillas are protected by law and up to 1974 capturers were strictly restricted in the exportation of gorillas and chimpanzees despite the highly attractive prices paid by the developed countries for these animals.

It was found that the methods of capture led to the loss of many of the apes. For example, to capture one young gorilla alive one needs to kill the mother, since there is always a battle between the hunters and the gorillas, which live in groups or in families. Faced with this situation, the government clamped down on animal capturers and since 1975 legal exportation of the apes has been reduced to rare and exceptional cases only.

The Government is also planning to strengthen law enforcement, in attempts to prevent poaching and illegal exportation of our wildlife species. The high prices paid for these animals in developed countries encourages these illegal practices.

In further efforts to conserve western gorillas, another area of 400 km^2 in the southeast near Mouloundou around Lake Lobeke, has been excised from timber concessions. This area is being reclassified for the conservation of elephants and other wildlife in general and probably gorillas in particular. Conservators now have been posted to Douala-Edea, Dja and Campo Reserves, the last two being important gorilla habitats.

(4) Research and information

Because our knowledge of the apes in Cameroon is very limited, we are prepared to receive investigators who will provide us with complete information. For example, in 1973 Miss Julie Webb from the University of California, USA, was authorised to carry out a preliminary study of the ecology and behaviour of the lowland gorilla in partial fulfilment of her M.Sc. degree.

She returned to the United Republic of Cameroon in January 1976 for a more detailed study of the gorilla. Her two year survey will include the Reserves of Dja, Campo, Takamanda and probably Pangar-Djerem.

Since 1973, many researchers have found their way into Cameroon. The FAO is developing management plans for the national park in the sahel north. This work should be completed soon. Mr. Doyle McKey of the University of Michigan, USA, carried out some studies of the ecology and behaviour of the black colobus (*Colobus satanus*) in the Douala-Edea Reserve for a Ph.D. degree. The final report is being awaited. Mr. Michael Kavanagh from Sussex University, England, worked on the green monkey (*Cercopithecus aethiops tantalus*) particularly in the north for a doctoral degree and we are awaiting his report. And of course, for many years Doctors Struhsaker and Gartlan have been involved in primate research in Cameroon. All these research activities are very important to us for policy making. As a result of the changes in our habitats, we have a number of endangered species such as the western gorilla (*Gorilla gorilla gorilla*), the drill (*Mandrillus Leucophaeus*), the black colobus (*Colobus satanus*) and the red colobus (*Colobus badious preussi*). Our policies must reflect up-to-date assessments of the status of each of these species.

PROBLEMS OF CONSERVATION IN CAMEROON

Cameroon is not without its conservation problems despite our untiring efforts. Our people love tradition, and their farming methods are traditionally based on shifting cultivation. This pattern of agriculture has affected our wildlife habitat and has facilitated illegal trapping and hunting. Agricultural development and nature conservation require co-ordinated planning, which in the United Republic of Cameroon comes under the auspices of the Ministry of Agriculture. Conservation, the wise use of natural resources, seems to us an essential part of the total development programme of our country.

Our basic need now is for conservation education, and some effort is being made in this direction. People are now familiar with the general principles and concepts. They still have to reconcile this new idea with their entrenched traditional ideas, that forests and wildlife are god-given and should be used as required, to satisfy man's needs. Thus the old traditional thinking had no room for conservation.

We now need conservation education in schools and at the village level. Much is being done in the north by the Director of the Wildlife College with Wildlife Clubs and in general by the Information Services of the United Republic of Cameroon. But we require new methods and new materials for such a task. There is dire need to establish and organise interpretive programmes in our conservation areas to enable local people and school children who visit and stay in the national parks to understand the importance of conservation. It is a sad fact that many Africans have seen African animals for the first time in the zoological gardens in Europe, the United States and Canada.

The conservation and protection of our savannah parks and reserves is well established. This is not to say that it is without problems, but principles of conservation are more easily practised in the savannah. The protection of rain-forest areas presents far more problems. We require basic scientific information based on field work before decisions can be made. In establishing a national park system in the rain forest, we need a lot of basic information before we can solve our management problems. This research will take time and cost money.

The actual protection of forested areas is more complex than in the savannah because of the difficulty in travelling and poor visibility. More game guards are required in the forest zone than in the savannah, but in practice there are fewer. The Government is quite aware of these problems and is doing all it can, with limited financial resources, to solve them.

The logistical problems are also often greater in forested areas. A major problem is that the bulk of our population lives in the forested zones, and there is a continual demand for dried and smoked game meat (bush beef) from our large urban centres. Poaching will not only continue but it has been increasing as a result of the fast growing population in these areas. Alternative protein sources need to be investigated. This is a serious problem because there are regions where cattle cannot thrive in order to solve the protein deficiency. Any suggestions on this will be welcome.

One other way to discourage illegal hunting would be to levy heavy taxes on cartridges and probably regulate their sale to the population. This problem, like all others already mentioned, is not insurmountable, but requires man-power, money, time and tact to avoid other social problems.

There are two specific conservation problems affecting Cameroon which should interest the International Primatological Society. The first of these concerns collecting wildlife for museums. We think that the collecting of mammals, particularly large and endangered species such as gorillas and colobus monkeys, can hardly be justified in the present day and age. Zoos and museums in Europe and the United States already have an abundance of such specimens. Zoos could encourage breeding in order to meet their needs. Dead animals in zoos in Europe and the developed countries could provide a sure source of material for museums. People requesting scientific permits for collecting specimens should ask themselves if these animals are really necessary. Perhaps in some cases specimens could be obtained from zoos already possessing such collections.

The second problem is even more serious and has occupied us in the United Republic of Cameroon for some time. It is the gorilla trade. Animal capturers are required to possess humane capturing equipment (tranquilizer guns, etc.) and to show proof that their animals were captured humanely. This has reduced the number of species such as the gorillas legally exported from Cameroon.

As the number of gorillas exported is reduced, however, demand exceeds supply and prices in Europe and elsewhere rise. Thus the temptation to smuggle the gorillas into Europe is greater. Since there is a real danger of gorillas becoming extinct in the wild, it is imperative that every effort be made to ensure that only legally exported animals find their way into the world market. People who find information indicating smuggling will assist our efforts in the conservation of wildlife by reporting this to the competent authorities. This is a subject which merits your views.

We in the United Republic of Cameroon are convinced that the wildlife in our country is our natural heritage which, if properly conserved and managed could become a source of foreign exchange through touristic attraction. It is an important base for research and education. Given time, knowledge and money, the few problems confronting our wildlife management will certainly be overcome.

ON HABITAT AND HOME RANGE IN EASTERN GORILLAS IN RELATION TO CONSERVATION

A.G. GOODALL

Department of Biology, Paisley College of Technology, High Street, Paisley, Strathclyde, Scotland.

The mountain gorilla (or Eastern gorilla of Groves, 1970) is listed as 'endangered' in the WWF Red Data Book. Seventeen years ago Emlen and Schaller conducted the only comprehensive distribution survey and census. They estimated the total population of mountain gorillas to be some 5,000-15,000. Further, they discovered gorillas were most unevenly distributed throughout their range. The animals were scattered in some sixty 'pockets' - many of which were more or less isolated. Conservationists may immediately recognise this as an early danger signal. The major threat to the continued existence of gorillas appears to be destruction of its habitat by man. This situation is clearly exemplified in the Virunga Volcanoes region and the plight of the gorillas there has been stressed by Schaller (1963), Dian Fossey and her many co-workers on Mount Visoke and more recently, with a most compelling ecological argument, by Spinage (1972).

However, there is a great danger of overgeneralization. Just as the behavioural ecology of the Virunga gorillas (or more precisely the Kabara/Visoke population) has been taken to be representative of *the* mountain gorilla, so too the proposed conservation measures will be presumed relevant to the survival of *all* eastern gorillas. The proposed strategy is conservation in its strictest sense - LEAVE WELL ALONE - NO INTERFERENCE. The point that I would like to make is that the relationship between man and gorillas is not so simple. It is in fact paradoxical that man may threaten gorillas with extinction by *destroying* their forest habitat yet man may also have contributed to the increase of gorilla populations in some areas and to their geographic spread into other regions. This he has brought about by a widespread practice in the past - not of forest destruction, but of vegetational *change*.

Such widespread changes in vegetational types can clearly be seen in the Kahuzi-Biega region of Zaire. In order to investigate the utilisation of these by gorillas I followed one group, of twenty animals, almost daily for seven months. They inhabited an area of mixed, montane rain forest near Tshibinda on the foothills of Mount Kahuzi at an altitude of 2,000-2,500 m. I found that the behavioural ecology of these gorillas differed markedly from the Kabara/Visoke population in many respects, including a far larger home range area (about 30 km^2 compared with up to 10 km^2), higher frequency of climbing, higher number of tree nests, habit of defaecating *outside* their night nest (and thereby *not* sleeping on their dung as the Virunga gorillas do), and far wider range of food items in their diet (over 100 cf. about 40). The most significant difference, however, was the markedly seasonal variation in utilisation of their home range.

The main proximate factor influencing most of these differences between the two gorilla populations is undoubtedly *the relative abundance, distribution and availability of suitable food items*. In this respect the forests of Kahuzi-Biega are far more variable than those in Virunga spatially, temporally and in their physiognomy. In particular, they have many more 'layers' and therefore are far more three-dimensional. Such differences are of course the consequence of the interplay of factors such as climate and geology, natural ecological succession and anthropogenic influences.

If one looks at vegetation maps of many areas inhabited by eastern gorillas today (such as the excellent ones prepared by Christine Marius/Weyns, University of Ghent) he sees a veritable 'mosaic' of many different vegetation types. These include primary rain forest characterised by many different dominant species in different regions, bamboo forest (*Arundinaria alpina*), hydroseres (*Cyperus swamps*), and many other successional stages of varying age and size structures from herbaceous to almost mature secondary and primary forest mixtures. This is clearly shown in the Kahuzi region and is more marked on the eastern side of the mountains which has been inhabited by the agricultural Shi tribes (who practise the 'slash-and-burn' techniques), than on the western slopes which have been inhabited by more forest hunters such as the Tembo.

George Schaller was the first to stress the importance of secondary regenerating forest in the ecology of eastern gorillas, but the situation, as demonstrated by the range utilisation patterns of the Kahuzi gorillas, is obviously more complex than a simple dichotomy between primary and secondary forests. I suggest that the very marked plasticity of gorilla structure and behavioural ecology has enabled them to exploit a veritable 'mosaic' of biotopes more fully (biomechanically, ecologically and nutritionally). It has also enabled them to extend their range and to colonise somewhat specialised areas such as the Virunga Volcanoes, where in fact many of the plant species present are typical of either disturbed areas or of secondary regeneration, e.g. the *Galium and Urtica* association, and even the *Hagenia abyssinica* woodland itself.

In considering the future survival of eastern gorillas, therefore, I suggest that it is the variety or 'mosaic' of different vegetation types and age structures which we must try to conserve. (There are also strong nutritional arguments for maintaining a variety.) Achieving this variety will require careful *management* of the forest in some areas and thus some primary forest *will* have to be felled - UNDER CAREFULLY CONTROLLED SUPERVISION. Such management practices could result in good public relations, for the timber would be most welcomed by the local people as a much needed source of fuel and house building - a point often overlooked by some ardent conservationists.

To summarise, I make the following urgent recommendations:

1. There must be strict enforcement of the existing statutes of National Parks, especially in Rwanda, (with more *international* help to ensure this).

2. A distributional census of eastern gorillas should be conducted in relation to vegetation types.

3. From the data supplied by such surveys the *most important* areas should be designated. President Mbuto Sese Seko of Zaire has recently announced that the National Parks of Zaire will be increased to cover 12-15% of the country. Since the majority of eastern gorillas are found within the borders of Zaire this is good news. Thus the survey work is doubly urgent to take advantage of this further appreciation by Zaire of its important role in habitat and wildlife conservation as soon as possible.

4. An overall management plan must be formulated for the conservation of *the* eastern gorilla. It is pointless to hope that *all* gorilla populations will be saved (and downright unethical where human lives may be at risk). Thus we may have to opt for a variety of different biotopes. I suggest that some of the gorillas inhabiting areas outside those designated for protection could be translocated into other suitable regions - the Ruwenzories for example. Some of the young gorillas should be sold to zoos and other institutions (*which have good facilities and good breeding records*) and the funds so raised then used to help finance the whole operation. This could eventually be a *two-way* movement, for any surplus animals resulting from successful breeding programmes in captivity (and these are beginning to appear for gorillas) could be returned to forest areas in Africa. Some of these areas may be in regions where forest has been allowed to regenerate. Alternatively such areas could be recolonised with animals translocated from another area which is about to be cleared of its forest. Thus in such a *rotational* management plan we could also rotate the gorillas (and other faunas) as well as the crops, timber, etc. whenever feasible.

5. More forest areas should be developed for tourism, like the successful Kahuzi-Biega National Park in Zaire.

6. There should be educational programmes, to inform the local inhabitants in simple terms about Wildlife Conservation, land use management, crop rotation, etc. (These should include slide shows like those being developed by Regina Frey and Monica Borner in Indonesia.)

7. To follow up Dr. Garine's suggestion, parallel nutritional studies of diets among gorilla and neighbouring human populations should be conducted - especially in areas with endemic protein deficiency diseases. It may be that humans can gain from eating some excess gorilla foods - though unfortunately this could lead to competition.

Finally, I feel that management programmes such as these, which try to incorporate aspects of wildlife conservation, rain forest conservation, economic gains via tourism, local support via provision of timber for fuel and building, and nutritional studies, (i.e. realistic land-use management schemes) will not only be more satisfactory in reducing the many conflicts of interest for rain forest land use - but in the present decade be more socially and economically relevant to the people who *actually live* in the so-called 'developing countries'. In his opening address, Dr. Bourne quoted, from a paper which I wrote with Colin Groves, a reference to 'the ultimate creeping threat' - this in fact did not refer directly to the destruction of the tropical rain forest but to the rapidly increasing human population. I suggest that our conservation and management plans should be geared to minimising this threat.

Stephen Rose, in his book 'The Conscious Brain', quoted Dennis Chitty who pointed out that 'few ethologists at a recent international conference got beyond doing for mankind what Beatrix Potter has done for the Flopsy Bunnies and Jemima Puddleduck; that is, describing the behaviour of one group of organisms in a language appropriate to another'. I hope we do not make the same mistake while formulating our plans for the conservation of primates.

REFERENCES

Groves, C.P. (1970). *J. Zool. Lond.*, 161, 287-300.
Schaller, G.B. (1963). "The Mountain Gorilla Ecology and Behavior". University of Chicago Press, Chicago.
Spinage, C.A. (1972). *Biol. Conserv.*, 4, 194-204.

PROBLEMS OF PRIMATE CONSERVATION IN A PATCHY
ENVIRONMENT ALONG THE LOWER TANA RIVER, KENYA

C. MARSH

*Department of Psychology, University of Bristol,
Bristol, England.*

Gallery forests along the lower Tana River in Eastern Kenya support 7 species of primates, including endangered subspecies of red colobus (*Colobus badius rufomitratus*) and crested mangabey (*Cercocebus galeritus galeritus*). The geographical ranges of these two populations coincide closely for 60 km between the villages of Wenje and Garsen. The total numbers are approximately 1,400 to 2,000 red colobus and 1,200 to 1,700 mangabeys. These estimates are based on visits to almost all forest patches in the range and data on mean group size from detailed studies of the mangabey (Homewood, 1976) and of the colobus by Marsh (in preparation). They broadly agree with estimates of Andrews et al. (1975).

The total area of occupied habitat is presently some 25 km^2, but it is broken up into over 70 separate patches, so that the average patch size is only 35 ha. Furthermore, analysis of aerial photographs has revealed a net loss of 15.9% of the forested area between 1961 and 1974. Some of this loss can be attributed to changes in the river course during the massive floods of 1961. The flood plain vegetation pattern is intrinsically dynamic and subject to a degree of fragmentation. In recent years, however, this has been greatly accelerated by the increasing human population and more intensive land use.

The Pokomo people practice shifting agriculture based on the irregular but potentially twice-annual floods, and traditionally clear new land at intervals. Meanwhile, the pastoral Somali and Galla tribes use the flood plain in the dry season, often burning the grass there to improve the grazing. This prevents forest regeneration and causes grassland to encroach upon forest edges.

Red colobus and mangabeys are not hunted locally, but the declining and fragmented nature of their habitat makes them vulnerable to local extinctions. One or the other species is absent from a number of patches which appear to be suitable habitat. For example, red colobus are absent from the largest single patch in their range (the 5 km^2 Wanje East patch) even though all their major food species are present. Local information and a careful examination of this forest suggests that it is mostly late secondary in character. It was probably cleared for farmland at least 100 years ago, causing a local extinction of one or both rare species. When the land was subsequently abandoned, it began to regenerate and eventually could support monkeys again. Mangabeys, with their semi-terrestrial habits and wide ranging pattern easily repopulate such areas, but red colobus rarely move more than 100 m along the ground and are much less likely to recolonize forest patches. However, they can survive in some very small patches from which the mangabeys are absent. This may be because these patches are too small for the mangabeys with their diverse dietary and ranging requirements (Homewood, 1976).

If these small patches of forest are considered as islands, local extinctions of primates accord well with predictions from the study of island biogeography of bird faunas (MacArthur and Wilson, 1967). Diamond (1975) has stipulated that extinctions can be minimized by establishing reserves with minimum perimeter to area ratios. In other words, a single large island makes a better refuge for wildlife than an equivalent area of small islands.

A Tana River Game Reserve has recently been created to conserve the best remaining habitat of the Tana red colobus and mangabeys (Marsh, 1976). If they are to survive it is essential not only that further forest destruction be halted, but also that fallowed land between patches be allowed to regenerate towards forest, so as to make a more continuous belt of habitat.

ACKNOWLEDGEMENT

I wish to thank the New York Zoological Society for financial support of this study during three years in the field.

REFERENCES

Andrews, P., Groves, C.P. and Horne, J.F.M. (1975). *J. E. Afr. Nat. Hist. Soc.*, 151, 1-31.
Diamond, J.M. (1975). *Biol. Conserv.*, 7(2), 129-147.
Homewood, K. (1976). "The Ecology and Behaviour of the Tana Mangabey". Unpublished. Ph.D. Thesis. University College, London.
MacArthur, R.M. and Wilson, E.O. (1967). "The Theory of Island Biogeography". Princeton University Press.
Marsh, C.W. (1976). A Management Plan for the Tana River Game Reserve. Report to the Kenya Game Department. 81 pp.

BIOECONOMIC REASONS FOR CONSERVING TROPICAL RAIN FORESTS

T.T. STRUHSAKER

New York Zoological Society, Rockefeller University.

INTRODUCTION

The rapid decimation of tropical rain forests and the short-sighted economic goals used to justify this destruction have prompted me to develop a rational and biologically-based argument for conserving large blocks of tropical rain forest. The argument has as its basis the economic and material well-being of mankind on a long-term scale, which I believe can be achieved only through the implementation of a land-use policy based on sound ecological principles.

The destruction of tropical rain forests stems from two basic problems: increasing human populations and increasing consumption per capita. Most tropical countries have human populations which are increasing at the rate of 2-4% annually with a population doubling time of 15 to 30 years (Ehrlich and Ehrlich, 1970). This means an ever-increasing demand for agricultural land and wood for local construction and fuel. The industrialized nations are actually increasing consumption per capita annually (anon. 1971). Most of this consumption is wasteful and senseless. Witness, for example, the tons of paper and plastics discarded daily in any Western country. The developing countries using the industrialized countries as models are aspiring to achieve similar levels of consumption and waste.

The rain forest is a highly complex habitat containing a greater diversity of plant and animal species than any other terrestrial ecosystem (e.g. Parsons and Cameron, 1974). However, its richness is deceptive. If the rain forest is destroyed for agricultural purposes, the nutrients which are tied up in the living forest are also destroyed. As a consequence, the land can be used for only short-term agriculture resulting in shifting cultivation and the continual destruction of more forest. Rain forests may take many thousands of years to regenerate, but destruction could be irreversible. The following sections outline important bioeconomic reasons for conserving tropical rain forests.

REASONS FOR COMPLETELY CONSERVING EXAMPLES OF MATURE RAIN FOREST

Natural heritage and tourism

These two points are commonly invoked in arguing the value of conservation and there seems little need to elaborate on them here. I would like to emphasize that tourism can be developed in rain forests, as exemplified in Zaire (MacKinnon, 1976) and Puerto Rico (Odum, 1970). The greater the variety of tourist attractions a country has to offer, the more successful will be its tourist industry. This in turn can mean more foreign exchange, an especially critical factor for developing countries. Some areas and habitats are less conducive to tourism than others and in some cases tourism may not be desirable. In these instances it is important to emphasize some of the other arguments developed below.

Watershed

The value of undisturbed forests as watersheds is perhaps one of the most obvious, important, and frequently cited reasons for conserving forests. The effectiveness of a watershed as a deterrent to rapid run-off of water, flooding and soil erosion is directly related to the extent of vegetation cover (e.g. Nye and Greenland, 1960; Bormann et al., 1968). Consequently, a rainforest offers the optimum in a watershed in contrast to alternative forms of plant cover, such as tea and coffee plantations. Watersheds are obviously of paramount importance to surrounding agriculture and fishing industries. In the absence of watersheds one can expect increased flooding and erosion and less year-round availability of water, particularly in areas of steep terrain.

Climate

Although the effects of forests on rainfall are difficult to study and the results are still considered ambiguous by some, empirical data summarized in Stanhill (1970) leads to the conclusion that forests have a positive effect on rainfall. This is supported by a computer simulation study in which Potter et al. (1975) show that massive deforestation in the tropics would lead to the following chain of events: 'increased surface albedo→reduced surface absorption of solar energy→surface cooling→reduced evaporation and sensible heat flux from the surface→reduced convective activity and rain-fall→reduced release of latent heat, weakened Hadley circulation, and cooling in the middle and upper tropical troposphere→increased tropical lapse rates→ increased precipitation in the latitude bands 5 to 25°N and 5 to 25°S and a decrease in the Equator-pole temperature gradient→reduced meridional transport of heat and moisture out of equatorial regions→global cooling and a decrease in precipitation between 45 and 85°N and at 40 and 60°S.'

Rain forests also constitute a major source of the world's oxygen supply (Odum, 1970). A given area of rain forest is considered to be 30 to 100 times more important to the stabilization of atmospheric gases than is an equal area of tropical sea (Odum, 1970).

Research standard

Examples of undisturbed and mature rain forest provide us with a standard which can be compared to neighboring lands exploited by man in one form or another. Such a control area is imperative for any meaningful scientific study of the effect of man's exploitive practices on the land, climate, hydrology and biology. In temperate forests of North America, for example, comparison of unfelled and clear-felled forest land revealed the adverse effects of felling on watershed and soil fertility (Bormann et al., 1968). Annual water runoff was 40% greater in the felled area than in the unfelled forest and in some months there was a difference of 418%. The soil in the felled area lost 57 kg/ha of nitrogen per year in comparison to the unfelled area which gained 4.5 kg/ha of nitrogen in the same period. Losses of other minerals (cations of calcium, magnesium, sodium and potassium) from soil of the felled area were as much as 20 times greater than from the unfelled area. The magnitude of these losses can be expected to vary with the degree of deforestation.

Another example demonstrating the value of maintaining control study areas of mature rain forest is provided by a study in Nigeria. Selective timbering had an adverse effect on the numbers of species and individuals of soil arthropods, with the greatest number and diversity of soil arthropods occurring in the undisturbed forest (Lasebikan, 1975). It is well established that soil arthropods play a major role in developing and maintaining optimal crumb structure and porosity in the upper 5 cms of soil which is most conducive to plant growth (Nye and Greenland, 1960). Consequently, any reduction in numbers and diversity of soil arthropods can be expected to have adverse effects on soil structure and plant growth.

A final example comes from the Kibale Forest of Uganda in which comparison of selectively felled and unfelled mature forest revealed gross differences in the primate populations. Selective felling resulted in reduction of most primate species, including the rare red colobus monkey. Selective felling followed by poisoning of undesirable trees had the greatest effect in reducing primate numbers (Struhsaker, 1975).

Genetic storehouse

Perhaps the most valuable aspect of tropical rain forests, and yet the one most difficult to explain to the non-biologist, is its untold wealth of genetic material. The great number of plant and animal species is only one indication of the vast genetic array. The considerable genetic variation between populations of many rain-forest species further increases this spectrum. The direct value to mankind of most of this genetic material is at present unknown. None the less, a considerable amount is known about ways in which rain-forest organisms can be put to direct use for mankind. For example, Dalziel (1937) summarizes a wide range of plant species from West African forests which are of direct benefit to mankind, many of them of a medicinal nature. A more recent and sophisticated analysis concerns the steroidal alkaloids in the Apocynaceae with obvious medical applications (Goutarel, 1964).

The main advantage of maintaining undisturbed blocks of rain forest is that they act as genetic banks or reservoirs which leave open a much greater number of options for development in agriculture, medicine, animal husbandry, pest control, chemistry, etc., than if these reserves were totally and irreversibly destroyed. A prime example concerns the use of various tree species for timber. In 1970 the tree species in the Kibale Forest that were considered of commercial value for timber were very few. The majority of species were classified as weeds and correspondingly were detroyed with arboricides. Within a few years the situation changed dramatically due to technological developments and increased demands for wood. The weeds of the past became commercially valuable, but this did not retrieve the vast amounts of timber destroyed by previous 'refinement' practices.

Timber exploitation is the least imaginative way of using tropical rain forests. With additional research many, if not the majority, of tree species may well prove to have greater values than that afforded by their timber.

Pest control

The rain forests harbour a number of species which prey on or parasitize insect pests of man, e.g. insectivorous bats, birds, primates, insects, etc.

One of the best examples concerns a study in Ghana which strongly suggests that forest arthropod populations are naturally regulated by fungal pathogens (Evans, 1974). Evans found that the incidence of lethal fungal infections among arthropods was inversely related to the degree of exploitation of the forest. Undisturbed forests with heavy shade and minimal seasonal changes had the highest levels of arthropods attacked by fungi throughout the year. In secondary or depleted forest the numbers of diseased arthropods were low for most of the year. These phenomena seem related to the maintenance of high humidity and relatively constant microclimatic conditions in the undisturbed forest, both of which are conducive to fungal growth. Arthropods commonly killed by fungi included ants and coccids. Both of these groups of insects are involved either directly or indirectly in diseases of cocoa. It is of interest that the fungal parasites of ants have never been found in cocoa plantations. Evans (1974) concludes by quoting Atkinson (1953): 'The nearer the conditions established by forest management approach those of the natural forest the less will be the danger of a breakdown of natural regulation, and the likelihood of damage (by insects) of an intolerable degree.'

Sustained yield of timber

There is an ever increasing demand for wood products, whether it be for paper and luxury items (such as veneers) in developed countries or for local construction material and fuel in developing nations. Clearly, we are rapidly approaching, if not already entrenched in, a wood crisis. This means that artificial softwood plantations will have to be expanded at a rapid rate, and that natural forests, particularly the tropical rain forests, will have to be managed with extreme care. If rain forests are to be managed in such a way as to produce a sustained yield of timber, then we must not only increase our knowledge of this complex ecosystem, but also completely conserve large blocks of undisturbed and mature rain forest as a means of insuring its regeneration in adjacent areas that are subject to selective and rotational exploitation. Conserving such blocks of undisturbed forest is necessary for several reasons.

1. The majority of rain forest tree species occur at extremely low densities, i.e. one adult specimen or less per hectare. Although much remains to be learned about the biology of most rain forest trees, it is clear that many of them do not achieve full maturity until 300 to 400 years old. In the few enlightened tropical countries practicing rotational timbering with the principle of maintaining a sustained yield of timber, the usual rotation cycle is about 70 years. This means that some trees will not have achieved sexual maturity before they are felled in the cycle and that virtually all emergent and commercially valuable species will not have produced their full complement of seeds by this time. Consequently, large areas of undisturbed forest must be conserved to insure a supply of seeds for proper regeneration in the adjacent areas undergoing exploitation.

2. Seed dispersal agents play a critical role in allowing plant seeds to overcome the potentially catastrophic effects of seed predation (Janzen, 1971b). In general, seed predation is greatest in the immediate proximity of the parent tree and where seed densities are high. Seeds transported by dispersal agents have a greater chance of survival because of increased distance from the parent tree and lower seed densities.

Many seed dispersal agents occur in low densities, such as hornbills, chimpanzees, elephants, fruit bats, and civet cats in Africa. Forest regeneration is dependent on these dispersal agents not only for alleviating the impact of seed predation, but also to ensure the spread of the tree species concerned. In most cases we are sufficiently ignorant of the life cycles of these dispersal agents as to be unable to predict the effect of deforestation on them. We do know, however, that some of them, such as many monkey species and chimpanzees (Struhsaker, 1975; Albrecht, 1976) decline in numbers in response to selective timbering. Consequently, it seems imperative that sufficiently large examples of undisturbed rain forest be maintained as a means of ensuring viable populations of dispersal agents which in turn are essential to regeneration in adjacent forests exposed to exploitation.

3. Most rain-forest trees are insect pollinated. In some cases the relation between tree and insect is very specific (Ehrlich, 1974). In order to maintain adequate populations of pollinators, as in the case of seed dispersal agents, it is essential to maintain blocks of their undisturbed habitat.

4. Natural regulation of insect populations, including those which attack commercially valuable trees, appears to be contingent on maintaining examples of undisturbed forest. Although the life cycles of most forest insect pests are poorly understood, the study of Evans (1974) cited above clearly demonstrates the great impact of fungal pathogens on insect populations and the dependence of these fungal pathogens on the optimal environmental conditions which are best provided by mature forest. Any major disturbance to the forest disturbs this regulatory system. I am suggesting that by conserving large areas of undisturbed forest we will also be conserving the natural regulators of forest insect pests.

In essence then, timber felling disrupts an extremely complex ecosystem which is in dynamic equilibrium. To ensure that the felled areas are capable of regenerating and thereby supplying timber on a sustained basis it is essential to maintain large neighbouring blocks of the undisturbed ecosystem.

HOW LARGE A FOREST FOR CONSERVATION?

In the preceding section it was pointed out that many rain-forest tree species, seed dispersal agents and pollinators occur in very low population densities. In order to maintain reproductively and genetically viable populations of these species large numbers of individuals must be conserved, which in turn means conserving large areas of forest. For large mammals it would appear that when a population falls to only a few hundred individuals their genetic diversity is essentially lost and consequently the population is less adaptable to environmental change than if it numbered in the thousands or more (e.g. Bonnell and Selander, 1974). Similar patterns may be expected for other groups of organisms, particularly k-selected species. One immediately thinks of the large home ranges and low population densities of seed dispersers such as chimpanzees and elephants, but many pollinating bats and insects also cover large areas. For example, in Costa Rica a euglossine bee species is a long-distance pollinator which can forage as far as 23 km away from its nest (Janzen, 1971a). These long-distance, low-density seed dispersers and pollinators are of particular importance to low-density tree species because they enhance genetic mixing and dispersal.

Large areas of forest must be conserved as a means of ensuring against natural disasters. Disease epidemics could readily destroy entire populations of important species if the park or reserve were too small. Wind destruction such as caused by hurricanes, cyclones or tornadoes, is another natural catastrophe which can have a profound effect on rain forests (Richards, 1952; Longman and Jenik, 1974). Again if the area conserved is too small, it can be totally destroyed with one major wind storm.

Terborgh (1974) and Diamond (1975) have clearly pointed out the significance of lessons from island biogeography to problems of establishing and managing rain-forest parks and reserves. Bird-species diversity on tropical islands indicate that extinction rates are faster on small islands than on large ones. For example, Barro Colorado Island, which is only 17 km^2 has lost 18 species of birds in 50 years. In contrast, Coiba Island (453 km^2) in the Caribbean is estimated to have lost only five species in 100 years, and Trinidad (4,828 km^2) only two species in the same period. Most of the species which have become extinct are those which occur at low densities even under the best of natural conditions and are usually large for their trophic class and/or are ground dwellers. Presumably, when populations of these species are too small they are more prone to extinction because of such things as natural variation in reproduction, resource availability, disease epidemics, etc. In an important study of mammals on mountaintops (a kind of terrestrial island) Brown (1971) demonstrates that, whereas birds are often able to recolonize islands because of their high mobility, most mammals are unable to do so, even when the islands are terrestrial. As in the ornithological studies, Brown found that the number of mammalian species inhabiting a montane island was closely correlated with the area of the island. He further points out that carnivores, mammals of large body size, and those with habitat specializations are most prone to extinction. Small, generalized herbivores survive best on terrestrial islands. These conclusions from biogeography clearly demonstrate the need of establishing parks and reserves of several thousand square kilometers if we are to balance the rates of extinction with the rates of evolution.

OPTIMAL DESIGN FOR FOREST CONSERVATION

Diamond (1975) in a very thought-provoking article, has outlined six design principles for terrestrial parks and reserves:
  1. A large reserve is better than a small one because it can contain more species and it will have lower extinction rates for these species.
  2. Division of the area must be avoided. A single large area is preferable to several small areas totalling an equivalent size. The same arguments apply here as in the first principle.
  3. If the area must be subdivided, then it is best to have these areas as close together as possible in order to increase immigration rates between the reserves. This will enhance recolonization between reserves of species gone extinct in any one of them.
  4. If there are several separate reserves, it is better that they be grouped equidistant from one another rather than in a linear arrangement. An equidistant arrangement allows greater opportunity for exchange and colonization between all of the reserves; exchange between reserves in a linear design is unlikely to be equal, particularly for the areas at either end of the chain.
  5. If there are several disjunct reserves, connecting them with corridors of habitat similar to that of the conserved areas will increase the chances of exchange and dispersal between the reserves at very little extra cost. This will be of especial importance to highly specialized species that are restricted to the habitat under protection.
  6. A reserve of circular shape is preferable to one of elongate shape because it minimises dispersal distances within the reserve. This reduces the chances of local extinction within the conserved area.

CONCLUSIONS

In terms of conserving particular species, it seems obvious that we must be concerned with their ecosystems and that we must broaden our approach to conservation issues. Trade restrictions are not enough to protect endangered species. They will not survive outside of collections unless their habitats are maintained. In developing our arguments for the conservation of rain forests or any other threatened ecosystem we must give more attention to biological findings and concepts from a wide range of disciplines. Only in this way can we develop compelling arguments which satisfy both those interested in aesthetics and those concerned with man's sustenance and accumulation of material wealth.

Furthermore, the concept of 'use' must be reconsidered. A block of protected rain forest is being *used* in all the ways described above even if man is not actively removing timber or other forest produce from it. These uses, though less tangible than in the conventional sense, are in many ways far more critical to man's well-being especially on a long-term basis.

It is clear that tropical rain forests constitute an essential part of the earth's life support system. For this reason alone we are compelled to demand that many large examples of this ecosystem be conserved throughout the tropics.

ACKNOWLEDGEMENTS

I am grateful to Lysa Leland for constructive criticism of the manuscript and secretarial assistance. Financial support was from NIH grant number MH 23008-04 and the New York Zoological Society.

REFERENCES

Albrecht, H. (1976). *Oryx*, 13, 357-361.
Anon. (1971). "Population growth: Boon? --or calamity?" Current Issues in Economics, American Telephone and Telegraph Company.
Atkinson, D.J. (1953). *Trans. 9th Int. Congr. Ent.* 2, Amsterdam, 1951, 220-223.
Bonnell, M.L. and Selander, R.K. (1974). *Science, New York*, 184, 908-909.
Bormann, F.H., Likens, G.E., Fisher, D.W. and Pierce, R.S. (1968). *Science, New York*, 159, 882-885.
Brown, J.H. (1971). *Amer. Nat.*, 105(945), 467-478.
Dalziel, J.M. (1937). "The Useful Plants of West Tropical Africa". Crown Agents for the Colonies, London.
Diamond, J.M. (1975). *Biol. Conserv.*, 7(2), 129-146.
Ehrlich, P.R. (1974). *Environmental Conserv.*, 1(1), 15-20.
Ehrlich, P.R. and Ehrlich, A.H. (1970). "Population, Resources, Environment". W.H. Freeman and Co., San Francisco.
Evans, H.C. (1974). *J. Applied Ecol.*, 11(1), 37-48.
Goutarel, R. (1964). "Les Alcaloides Stéroidiques des Apocynacées". Hermann, Paris.
Janzen, D.H. (1971a). *Science, New York*, 171, 203-205.
Janzen, D.H. (1971b). *Ann. Rev. Ecol. and Syst.*, 2, 465-492.
Lasebikan, B.A. (1975). *Biotropica*, 7, 84-89.
Longman, K.A. and Jenik, J. (1974). "Tropical Forest and its Environment". Longman, London.

MacKinnon, J. (1976). *Oryx*, 13(4), 372-382.
Nye, P.H. and Greenland, D.J. (1960). "The soil under shifting cultivation". Technical communication no. 51, Commonwealth Bureau of Soils, Commonwealth Agricultural Bureaux, Farnham Royal, Buckinghamshire, England.
Odum, H.T. (1970). ed. "A Tropical Rain Forest: A Study of Irradiation and Ecology at El Verde, Puerto Rico". Atomic Energy Commission, USA.
Parsons, R.F. and Cameron, D.G. (1974). *Biotropica*, 6(3), 202-203.
Potter, G.L., Ellsaesser, H.W., MacCracken, M.C. and Luther, F.M. (1975). *Nature, London*, 258, 697-698.
Richards, P.W. (1952). "The Tropical Rain Forest". Cambridge University Press, Cambridge.
Stanhill, G. (1970). *In* "Analysis of Temperate Forest Ecosystems" (D.E. Reichle, ed.), Springer-Verlag, Berlin.
Struhsaker, T.T. (1975). "The Red Colobus Monkey". University of Chicago Press, Chicago.
Terborgh, J. (1974). *Bioscience*, 24(12), 715-722.

THE VIEWPOINT OF A CONSERVATIONIST

R.S.R. FITTER

Fauna Preservation Society, c/o Zoological Society of London, Regent's Park, London NW1 4RY, UK.

Some excellent suggestions have been made by previous speakers on the problems of primate conservation. They have, in my view, only one defect. They are addressed to a world situation which no longer exists. At some time in the past fifteen or twenty years, unrealised by most conservationists, we have passed from a situation in which the ecosystems of the world, the biosphere if you like, was deteriorating at a rate of arithmetic progression to one in which it is palpably deteriorating in geometric progression. Ideas and methods that would have been adequate for the former situation - the creation of national parks and reserves, autecological studies and setting up of captive breeding stocks - are quite incapable of coping, by themselves, with the rate at which the rain forests, for instance, are disappearing and the oceans are being polluted, so that there are now serious forecasts of the possible collapse of their ecosystems.

To be brutal, we are fiddling while Rome burns. There is only one way in which we shall be able to conserve adequate samples of the world's ecosystems, the habitats of primates and all other animals, in the 21st century - and the pocket-handkerchief-sized areas constituted by most national parks and reserves are decidedly not adequate in the long term. That way is to make the peoples of the world *want* to conserve these ecosystems. In democracies certainly, even under dictatorships ultimately, things only happen if the majority of the people want them to happen. At the moment there is hardly any country in the world - Papua-New Guinea is one of the handful - where the people's representatives understand the importance of conserving what is left of their natural or semi-natural ecosystems. Somehow we must convince a much wider spectrum of the people of all countries, and especially in the Third World, of the vital importance, for their own future, of not destroying any more wild habitat than is absolutely necessary and certainly of not destroying it in the present haphazard and unplanned way.

Most people are perforce primarily interested in the basics of existence, adequate food and shelter and good health. They are not likely to be impressed by advice from foreign experts, however distinguished, which does not recognise these primary needs. The task before us is therefore an extremely difficult one. We must find a way of raising world living standards on a long-term basis without destroying their natural heritage, which would inevitably lead to a rapid fall in those living standards. Unless we can do this, people everywhere will vote for or support the present short-term destructive policies which seem to lead to better things, but are in fact destroying the seed corn. If we confine our effort to saving a few square kilometres of rain forest here and there, or taking a dozen rare primates into captivity to find out how to breed them, we shall lose the whole shooting match.

SUMMARY REMARKS ON PRIMATE CONSERVATION

RICHARD W. THORINGTON, JR.

*Curator of Mammals, Smithsonian Institution,
Washington, DC 20560 USA.*

During the symposium, there were numerous exchanges of ideas and attitudes, both optimistic and pessimistic. Among these were certain recurring themes, involving what help is needed to effect the conservation of primates. A repeated answer is to be found in various forms in the papers by Asibey, Awunti, Brotoisworo, Mohnot, and Vanzolini (this volume). In short, it is to provide the scientific bases for conservation and wildlife management. Time and again it became clear that we do not know enough about the non-human primates to conserve them. Dr. Vanzolini noted that we must make the right decisions the first time, for we shall not have a second chance. Reserves must be correctly placed, they must be large enough, and they must be wisely managed. Dr. Awunti and others noted their needs for advice on managing wildlife. Dr. Asibey pointed out the need to experiment with the synchronous management of forests for timber and wildlife. If forests can be managed simultaneously for the two, then very different and aggressive conservation policies can be promulgated.

Although it is most obvious that conservation is dependent on ecology, other disciplines are equally essential. Taxonomy is crucially important. Without a good taxonomic understanding of primates we cannot determine which species are endangered. Taxonomists should be haunted by two conflicting nightmares. First is the spectre of excessive splitting which causes conservationists to expend unnecessary effort to protect invalid species. Second is the nightmare of discovering interesting cryptic species of primates too late - after they have become extinct. How many sibling species are concealed among the primates? How many cases of clinical variation in morphology conceal genetic discontinuities?

Morphology, physiology, and similar disciplines can contribute to important decisions in conservation. It will probably be impossible to save all of the species of the *Cercopithecus* radiation, or all of the *Saguinus*, or all of the *Presbytis*. It is most desirable that hard decisions about which species to save will be based on the biology of the animals, rather than on the politics of the situation. Can the morphologists and physiologists designate which species are the most unusual or the most typical of each genus? Noting Vanzolini's recommendation to preserve processes, should we place especial effort on conserving certain species because they appear to be undergoing rapid or divergent evolution in morphology, karyotypes, physiology, etc.? Perhaps the greatest burden falls on the medical profession, which must decide now which species to breed in captivity for medical research. The decisions will soon be irrevocable. Like the timber companies described by Asibey, the medical profession should be conducting salvage operations to obtain their breeding stock of especially selected species of monkeys from areas destined to become agricultural. The options are being reduced annually. If captive colonies of langurs or spider monkeys will ever be needed for medical research, they must be obtained now.

There is a great need for genetic research of relevance to conservation. We must be concerned about the genetics of small populations. One of the rationales for large reserves is the maintenance of genetic heterogeneity in the populations of large birds, mammals, and rare species. Yet we do not know how much genetic diversity there is in most populations of primates. The howler monkeys of Barro Colorado Island passed through a population 'bottleneck' in the late 1940's to perhaps as few as 200 animals. There are now ten times that many monkeys on the island. Will this apparently prolific and successful population have 'genetic problems' in the future, or is a forested reserve of 1,500 hectares adequate to preserve this species in perpetuity? How much genetic heterogeneity is there in small isolated populations of primates in Africa, Asia, and South America? Because some of these have been isolated on mountains or islands for thousands of years, studies of the population sizes and genetic homogeneity would be extremely relevant to our evaluations of the long-term success of various reserves and parks. What genetic problems, if any, are evident in situations like the small islands of habitat inhabited by the mangabeys and *Colobus* of the Tana River, as described by Marsh?

Ecological research seems not to keep pace with the needs of conservation. The conclusions of Muckenhirn and Eisenberg exemplify the importance of the most basic ecological and geographical reconaissance. We need to know where the different species of primates occur and, as noted for Brazil and Guyana, we are frequently ignorant. We need to know in which habitats they reach their greatest densities. These are some of the essential data requested by Vanzolini, Awunti, Asibey, and Brotoisworo, because they are needed prior to the establishment of reserves and required if wise policies regarding management are to be made. For example, Barro Colorado Island is being 'managed' according to the popular strategy of non-interference. This is clearly a good management policy for the maintenance of the howler monkey (*Alouatta palliata*) and probably it is also best for spider monkeys (*Ateles geoffroyi*). Yet if the objective were to maintain diversity on the island, or to maintain the populations of *Saguinus* and *Cebus*, it is probable that a different management policy, involving selected clearing and the maintenance of more early second-growth habitat, would be preferable. Similarly, Goodall suggests that the same policy of maintaining a mosaic of habitats is the best for managing gorilla populations. This hypothesis will not be greeted with universal approval or enthusiasm. However, it behoves primate ecologists to test this hypothesis as quickly as possible against distributional data of the plants and animals, because the wildlife departments of the relevant African countries need to make their decisions about gorilla management now. Marsh's conclusion that *Colobus* lack only the mobility to repopulate small forests along the Tana River has an immediate implication for the management of this species. The techniques are available and tested to capture and transfer *Colobus* to forests they don't presently inhabit. In such cases it would seem desirable to wed conservation, ecology, and population genetics into an experimental management program.

As discussed by Struhsaker, biogeographers are applying their knowledge to conservation and making suggestions about the size and shape of reserves. There is an ongoing debate, as demonstrated by the exchanges between Simberloff and Abele (1976a, b), Diamond (1976), Terborgh (1976), and Whitcomb et al. (1976). The main debate is whether we should always opt for large reserves instead of a number of small ones, but there are a number of other disagreements or different viewpoints expressed by the authors. It is questionable whether some of the issues are really relevant to primate conservation. For example, Diamond's concerns about 'weed species' or 'fugitive species' are probably irrelevant to primates. Other issues are demonstrably relevant. Simberloff and Abele (1976a) contend that small noncontiguous reserves may protect animals from catastrophes that could sweep through a large reserve. These conditions have been simulated by the recent isolation of the howler monkeys of the Darien from those of the Canal Zone by the transisthmian highway in the Republic of Panama. The yellow fever epidemic of the late 1940's affected all of Central America, whereas more recent epidemics have hit only the Darien howlers and have not decimated the *Alouatta* of the Canal Zone and further north.

There are other reasons to question the universal validity of the large reserve hypothesis. First, the equilibrium theory of island biogeography becomes largely irrelevant for primates if we are willing to reintroduce a species if it becomes extinct. I propose that this should be tried regularly in cases like that of the Tana River *Colobus*. Second, the equilibrium theory is a numbers game and is not concerned with the mechanisms involved. Wildlife management is or should be very concerned with mechanisms. Species do become extinct on islands like Barro Colorado, but many of these extinctions could be prevented by different management strategies. It will be easiest to manage reserves for single species, harder for several species, and it will require great wisdom to manage them for long-term maintenance of the greatest diversity.

Of course there are valid reasons for and definite circumstances under which large reserves are to be preferred. Many of these are summarized by Struhsaker and in the papers he cites. The arguments are most persuasive for large carnivores and other rare species. However, there is need for a 'mixed strategy' of large and small reserves for most species. I return to Vanzolini's theme that we should be preserving biological processes. We should think of our biological reserves as the refugia of the Recent Epoch, analogous to the refugial areas of the Pleistocene. We should expect evolutionary processes to be different in large and small reserves. In essence, we are establishing the evolutionary experiments of the next few centuries. We should continue to argue about the desirable sizes of reserves, but we should also be collecting both prospective and retrospective data and testing our theories with our observations.

These are a few of the ways in which different scientists can contribute to conservation. There are countless other ways as well. The other major side of conservation is the economic and political side. It provides as many opportunities and as many problems as the scientific side, but primatologists tend to be less capable of dealing with them. Some of these problems were discussed at the symposium and were presented in the papers. The ultimate problems of population growth and increasing use of resources loom impossibly large. Yet in the very presence and presentation of our speakers from South America, Africa, and Asia, there is hope and encouragement. Brotoisworo proposed to use the intricacies of Adat law in Indonesia. Asibey proposed to test the tourist potential of high forest primates. Awunti documented the willingness of the Republic of Cameroon to establish large reserves. Vanzolini outlined how to use an opportunistic approach to conservation in Brazil. We cannot afford to be too optimistic or to make unrealistically sanguine projections. But neither can we afford to become so discouraged as to quit or so dismayed with the prospects as to leave the work for others to do. If we wish to maintain a vigorous field of primatology, we must all co-operate to conserve primates. We do not need to agree, to do so would be stifling, but we must work together. We must make the bases of primate conservation as sound as possible, and we must stand ready to help our colleagues in tropical countries prevent the wholesale extinction of our fellow primates.

REFERENCES

Diamond, J.M. (1976). *Science, New York*, 193, 1027-1029.
Simberloff, D.S. and Abele, L.G. (1976a). *Science, New York*, 191, 285-286.
Simberloff, D.S. and Abele, L.G. (1976b). *Science, New York*, 193, 1032.
Terborgh, J. (1976). *Science, New York*, 193, 1029-1030.
Whitcomb, R.F., Lynch, J.F., Opler, P.A. and Robbins, C.S. (1976). *Science, New York*, 193, 1030-1032.

ROUND-TABLE DISCUSSION ON REHABILITATION

Summarised by MONICA BORNER* and PAUL GITTINS**

*Orang-utan Rehabilitation Centre, Bohorok, Sumatra.
**Sub-Department of Veterinary Anatomy, Cambridge.

INTRODUCTION

   Although the title of this session was 'Rehabilitation' several other important conservation topics were discussed. Rehabilitation is training young animals with little experience of the wild or of conspecifics, to live a natural life. This must be distinguished from the translocation or resettlement of newly-caught wild animals or the establishment of controlled breeding colonies. Rehabilitation and resettlement are both concerned with the introduction of animals into their natural habitat and as such are concerned with the protection of that habitat. The establishment of breeding colonies is usually concerned only with the preservation of the particular species. Most of the discussion centred on the rehabilitation of orang-utans since most work has been done on this species. In Senegal, however, a rehabilitation station for chimpanzees has been established and the initial results of this project were reported.

REHABILITATION

   The discussion was opened by Regina Frey who outlined the methods and achievements of the Bohorok Orang-utan Rehabilitation Station.
   The station was set up in 1972 at the edge of the West Langkat Reserve in North Sumatra where wild orang-utans occur. Animals for rehabilitation came from two main sources: pets that had been in captivity for several years were either confiscated with the co-operation of the Game Department or donated by their owners as the aims of the station became known. In two instances, wild animals living in areas about to be logged were rescued and resettled immediately deep in the reserve. Rehabilitation procedures were developed to help the animals become accustomed to the forest.
   After stays of varying length, animals were released deep in the Langkat Reserve to ensure they did not return to the station or come too close to human interference. In three years about 30 animals were introduced into the 2,000 sq km reserve. Initially it was thought essential to release the animals into areas where there were wild populations of orang-utan to guarantee that the habitat was suitable. It has subsequently become clear, however, that it is irresponsible to go on releasing animals into a restricted area without knowing what effects this has on the wild population's numerical balance with the habitat and because of the possibility of introducing disease into the wild population. Therefore it is planned in future to release animals into areas of North Sumatra where no wild orang-utans occur today but

where they occurred until recently, and to protect these areas from further human incursion.

What are the positive effects of rehabilitation? The orang-utan is a seriously endangered species. There are estimates of the population in North Sumatra ranging from about three to ten thousand animals. It is obvious that one cannot save this population by releasing some thirty animals. Today orang-utans are not primarily endangered by poaching, but by the destruction of their habitat caused by logging, oil exploitation and agricultural activities. The most important task of a rehabilitation station must be propaganda and conservation education. An increasing number of tourists, local and foreign, visit the project which offers a splendid opportunity for conservation education. The stations provide a clear example of wild life conservation in action, often so difficult for local people to comprehend. In addition, by providing a place for confiscated animals which for obvious reasons should not be killed or displayed in zoos, stations encourage local conservation officials to enforce the game laws. This has led to a great decline in the number of orang-utans illegally captive in or exported from North Sumatra.

In the course of the discussion, Regina Frey and Monica Borner were asked to explain in more detail the actual procedures involved in rehabilitation of orang-utans.

New arrivals at the station were put through quarantine for at least four weeks, in which time they were vaccinated, treated for worms and blood and faeces were taken for laboratory tests. Then they were moved to a second cage in the forest to get used to their surroundings. They were released and fed on top of that cage twice a day. Initially, milk and bananas were given which are important for young or undernourished animals. But as soon as they were released, all the animals started eating leaves and tried almost everything that looked edible. They never poisoned themselves although they would sometimes bite into something and spit it out immediately. The additional food was purposely kept monotonous, to encourage the orang-utans to look for food themselves. When the animals had been at the station for some time, the quantity of food was reduced and finally stopped. The ideal age for this seems to be around 4 years when the animal would naturally become independent of its mother. Monitoring the weight of individuals showed if they were able to fend for themselves. Note was made of the animals' climbing ability, ability to build and use nests and social behaviour when meeting wild orangs such as mating and touching the babies of wild mothers. When food was reduced some animals left the station, others were taken to the centre of the reserve and released there. Some of the animals who left the station of their own accord came back after several months, showing they had survived at least this long. Of about 60 animals received in three years, eight died, 30 have been released, the rest are still at the station.

One of the main points of the discussion centred on the value and function of rehabilitation centres today. Regina Frey in her opening statement suggested that the most important function of the rehabilitation centres is education. Herman Rijksen returns to this point and states that the actual success of such projects is that the government, pressed by outside opinion, starts to appreciate the situation. Today they are willing to remove timber concession out of the reserves. The Brindamours have worked hard especially in that direction in Kalimantan and are able at the moment to preserve their reserve. John MacKinnon went on to expand the point. Originally, rehabilitation stations were used as a means of redressing poaching deficits. Today, the conservation of the habitat has become their main duty. Small bits of forest with an overpopulation of orang-utans are no solution. Also to avoid the introduction of sick animals into the wild population, other areas should be opened for orang rehabilitants. The main aim of the stations is not really to reintroduce animals in the wild, but to facilitate the enactment of the conservation laws. The focal points for the stations are today: law enforcement, propaganda, education and possibly, tourism and research. The rehabilitation projects therefore still have a function today, but they are themselves not a solution to the deforestation problem. It might be dangerous if people give funds only for rehabilitation projects and think that they save the orang-utans while the main problem remains the destruction of the forest. Most of the participants in the discussion agreed with these points.

Alan Goodall asked whether there was any information on rehabilitation of other species, and asked for advice on how gorillas might be rehabilitated. Monica Borner stated that orang-utans seem to be especially suitable because they do not have to be introduced into pre-existing groups. Herman Rijksen had earlier pointed out that it is harder for male orang-utan rehabilitants to make their way within the wild population. Releasing them into an area without wild orang-utans might give them a better chance to reproduce. Mentawai gibbons have been released in groups of two or three as single animals could not form groups and would remain near the releaser's house.

Stella Brewer and Rafaella Savinelli then informed the meeting that they ran a rehabilitation station for chimpanzees in Senegal. They stated that because chimpanzees were social animals it seems quite easy to release them in groups, the young animals imitated the older ones and learnt quickly.

The station in Senegal began with a six-year-old female and two five-year-old males, who had come straight from the wild without being in captivity. These helped rehabilitate the other chimps. Long walks were made in the vicinity of the camp, so the chimps got to know the area and wild fruit trees. Usually, they would not hesitate to learn, by imitating a 'mother' until they were 7 or 8 years old, when they would start leaving their 'mother' naturally. At this age they also started to leave the station. It is important to give them affection and security in a 'family' up to that age.

Many of the edible fruit were known in the vicinity and these were shown to the chimps. On one occasion, however, the chimps refused to eat a fruit similar to a known edible fruit from Gambia. This proved later to cause violent sickness. The chimpanzees had to be taught how to build nests in the trees. One female didn't learn and had to be frightened at night by the workers wearing masks. Herman Rijksen added that orang-utans react instinctively to predators such as snakes displaying and shaking branches. Two chimpanzees, a male and female, were now completely independent, and had bred and produced an infant. There were six more chimpanzees in camp.

In the experience of the workers at Bohorok, in the rehabilitation of orang-utans, it seemed unnecessary to teach them what foods to eat or how to build nests, although some learning from more experienced animals has occurred. It would seem from the example given by Stella Brewer that chimpanzees also 'know' what foods are edible, but that teaching is necessary. In both cases no good data are presented for how much learning occurs or how, so no conclusions can be drawn. Paul Gittins pointed out that it does seem likely that chimpanzees living in large social groups could learn more by imitation of group members than orang-utans, a solitary animal, who must learn more by individual trial and error or have more innate behaviour patterns.

RESETTLEMENT (Translocation)

Alan Goodall returned to the point of resettlement of animals and stated that this could be an answer to the threat to the mountain gorilla. The secondary vegetation growing over slash-and-burn areas seems to produce an improved habitat for gorillas. He asked if selective logging had the same effect for orang-utans. John MacKinnon answered that orang-utans being arboreal, need the complete forest structure for locomotion. It should be possible to work out logging schemes, where selectively logged areas would remain untouched long enough to let the canopy recover and the orang-utans move in. Monica Borner added that it was not very difficult to catch and move orang-utans and that she had done it several times. It is very expensive, however. It would probably be easier and have the same effect, if a logging scheme could be worked out, where regenerated areas are adjacent to undisturbed habitat so that orang-utans could recolonise naturally. Isolated areas could be repopulated artificially with animals translocated from areas being felled.

Alan Goodall concluded by saying that 'land-use' was a vital point to be considered and that overall land management schemes must be prepared.

ISLANDS

Mrs. Neago-Biroum made a proposal at the start of the discussion to keep as many orang-utans as possible under semi-wild controlled conditions on islands outside southeast Asia. She stated that orang-utans should be protected outside rather than inside their countries of origin where the rain-forest, their natural habitat, is being destroyed. Rehabilitation stations have been in existence for a long time, yet orang-utans are still dying out. Trying to convince the local people to conserve their wild life does not work in Indonesia where they will not be told by foreigners what to do.

A controlled breeding colony of 7 chimpanzees has been established on an island in South Carolina with NIH funds, and now it is planned to extend this on a larger scale with chimpanzees, gorillas and orang-utans. It is hoped to obtain an Australian island near the equator on a 99-year lease.

It was immediately pointed out that there was a difference between the preservation of one species and nature conservation and that it is a priority to conserve animals in their natural environment. Alan Goodall commented that it is too patronizing to say that Indonesians are not able to look after their own animals or forest and that our experience gives us the right to put those animals on islands outside Indonesia where they would be kept by us! It is much more important to educate the people so that they will be able to protect the animals themselves.

On agreeing that islands were *one* possible method of preserving the orang-utan the discussion now considered in more detail the merits of an island as opposed to alternative sites.

If an island could be found with suitable vegetation such an experiment using surplus animals from zoos could be attempted. An island with forest suitable to wild orang-utans, however, would almost certainly be subject to the same deforestation problems as other areas. Paul Gittins pointed out that if orang-utans were fed, or even if fruit trees were planted, the situation would never be natural. These animals would still have to be fed and looked after in 20 years time, whereas other projects like rehabilitation or resettlement of animals within natural habitats could be left alone and would stabilise themselves. The cost of transport of animals, providing food and the necessities of permanent maintenance would make island projects veyy expensive. If money is available to buy islands, it may be better to buy land in existing orang-utan habitat, despite the fact that problems of control of animals and poaching may be greater.

SUMMARY

1. Although the rehabilitation of orang-utans is possible and methods of achieving this are now established, the main emphasis of rehabilitation stations has changed. Today the most pressing problem facing the orang-utan is the destruction of its habitat and the function of rehabilitation stations must now be to try to check this through propaganda and conservation education.
2. Because of the danger of upsetting the balance of wild populations, animals must in future be released into areas where orang-utans are no longer found. Areas close to their present range should be chosen first, and also experiments of releasing animals into selectively logged areas should be made.
3. Resettlement of endangered populations should be attempted, but a better method would be to log the forest lightly in slow rotation so that the wild populations could move unaided from areas being felled into areas where sufficient regeneration has taken place.
4. Islands could be set up as semi-wild breeding colonies utilizing surplus orang-utans and thus ensuring the continuation of the species for possible reintroduction into the wild at some future date.
5. Rehabilitation is possible with other species; chimpanzees are now being rehabilitated in Senegal. It seems, however, that the social structure of the species to be rehabilitated must be considered and different methods developed to suit each. In social species it may be necessary to release animals in groups to overcome the difficulty of introducing single animals into existing wild groups.

SECTION II

TRADE AND SUPPLY OF PRIMATES
(R.E. Hackett, ed.)

TRADE AND SUPPLY

R.E. HACKETT

Shamrock Farms (GB) Ltd, Henfield, Sussex, UK.

It is likely that most of the current trade in non-human primates is for the purpose of supplying the needs of zoos and the scientific community.

In some countries, the pet trade still flourishes, although reduced by tighter controls (e.g. the UK Rabies Order) and by dwindling supplies. This latter problem is dealt with elsewhere: our concern is with such matters as acquisition (with survival of the species in mind), transportation and economics.

The supply situation with regard to some species gives rise to anxiety, although it is more properly controlled now than ever before. Embargoes by governments, climatic conditions, in some countries the attitude that primates are vermin, and genuine depletion of natural populations through destruction of habitat and over-trapping in some areas, have contributed to an unsatisfactory situation where, on the one hand, well-meaning people would ban the export of everything (including the vermin) and on the other hand where sincere efforts are made to protect endangered species and to avoid others becoming endangered, whilst still ensuring a supply for the purposes mentioned above.

Perhaps the greatest hazard to the achievement of a satisfactory solution is the existence of certain unscrupulous persons at each end of the line, such as those who smuggle animals out of countries of origin (as for instance, in certain South American countries where embargoes exist), the dealers (as opposed to the genuine suppliers) who will buy them, and the research worker who will use them, knowing that they have been obtained illegally.

Animals supplied in this way invariably suffer heavy losses, due often to inadequate care during a journey which may necessitate transit through a number of countries, to avoid the attention of authorities who are conscientious and cannot be bribed. This, of course, affects the economic aspect of the situation as the cost of the lost animals must be added to that of the remainder.

Transport problems have largely been overcome thanks to the concerted efforts of the authorities in both countries of supply and countries of destination, the genuine suppliers and the airlines.

We have tried in this session to consider as many as possible in a short time, of the aspects connected with the trade and supply of sub-human primates and we hope that as a result the true facts are now more clearly in the minds of many people, who hopefully may contribute something towards the solution of the problems still in existence.

THE SUPPLY OF MONKEYS FROM PENINSULAR MALAYSIA

A.C. LAURSEN

583 Jalan 17/17, Petaling Jaya, Selangor, Malaysia.

INTRODUCTION

This paper is intended to give research primatologists an idea of the many influences exerted on the monkey between its wild state in Malaya and its final arrival in his laboratory.

THE TRAPPER

In Malaysia the trapper is an amateur in the sense that trapping is a sideline to supplement income, rather than a full-time occupation. Monkeys (*Macaca fascicularis*) have been quick to realise living is easier in the cultivated and built-up areas, and have deserted the less accessible deep jungle and forests for the developing areas. The present day trapper, therefore, is a local man whose regular occupation permits some spare time to set traps and collect the 'catch' later. Most trapping is done on private property or on land gained access to through private property and any trapping by outsiders or teams of full-time trappers would encounter difficulties.

The trapper is in business purely for economic reasons. Often he has little regard for animals in general, and looks upon the monkey in particular as a destructive pest. On the other hand, he has a high regard for money and may tend to spend the absolute minimum on care and maintenance of his catch. Hence many of the problems frequently experienced with these monkeys are initiated in the period following capture. Liaison between the trapper and the exporter is vital and in Malaysia this is quite good. The system functions, not through educational propaganda, as advocated by some academics, but with a little economic blackmail. Unfortunately, in times of high demand for stock, the weapon occasionally becomes ineffective, as the trapper is well aware that if one exporter will not buy his animals because of their poor condition, another purchaser can readily be found.

THE COLLECTOR

The collector may be an employee of the exporter, or an independent entrepreneur. Owing to the distances and points of the compass involved, the number of trappers to be visited (most with small catches, and some without any) the exporter can ill afford to do the job himself. The constant daily managerial supervision required by the holding unit would otherwise be neglected. Whether or not an employee is used as a collector depends a good deal on how much reliance the exporter places on the integrity of his employee because purchases are made in cash.

With certain reservations, the use of an independent collector is perhaps the best method of obtaining monkeys. The exporter can leave him to make contact with and organise local men to trap in a given area and to help and advise those new to the job. He will purchase and collect from each of his individual trappers. When a significant number has been collected he will deliver this batch to the exporter and will be paid cash on a per capita basis. The collector is expected by the trapper to buy *all* the monkeys trapped, regardless of size and quality - although a slightly lower flat rate may be negotiated between them if the quality is obviously below standard, or the size distribution is peculiar. On the other hand he ought to ensure that, if the bargain is to buy all the catch, the more popular sizes have not been removed by some other buyer.

The trapper probably captured monkeys from the same family group but the collector is obliged to hold in a small area, and transport together, monkeys from many different groups in his district. If the monkeys are mixed indiscriminately, the establishment of a new 'pecking order' may have disastrous results.

Because the collector has direct contact with the exporter, he is much more aware of the importance of supplying good quality stock than is the trapper. He is the crucial link in establishing communication, whether in respect of numbers required or quality of catch, between the exporter and the trapper.

THE EXPORTER

It is the responsibility of the exporter to ensure that the primates ordered are of top quality and are forwarded under the very best conditions available. The management, husbandry and the veterinary attention he supplies in his holding unit will vitally affect the health and quality of the primates received by the customer. A good exporter will personally select each animal on the day of packing for export, rather than leave this hot, dirty and fatiguing chore to others and will choose a carrier whose flight schedule and route will best suit the welfare of the monkeys. In order to ensure a smooth flow of exports, it is important that he should co-operate with and be well regarded by the Game, Veterinary and Customs Departments.

THE CARRIER

Having carefully selected and suitably packed the catch, confident in his own mind of its good health and high quality, the exporter passes it on to the carrier. The local cargo managers make every attempt to ensure the welfare of animals in their care while awaiting embarkation, but once the shipment is air-borne the treatment meted out to the monkeys is beyond their control. Good managers do advise colleagues along the route that livestock is aboard and request attention be paid to their needs. They also confirm to the consignee that the shipment is en route and thus help ensure that delays in clearing the shipment at its destination are minimal. However good the local airline staff, there is no insurance against hazards including last minute changes of schedule, overflying scheduled stops, mechanical failures, staff strikes and indifferent airport staffs en route. Usually all goes well, however, and the cargo arrives at its destination in good condition, but on many occasions an impending disaster has been averted by effective and rapid communication.

SUMMARY

The chain of supply of monkeys (*Macaca fascicularis*) from Malaya includes the trapper, who catches animals in a relatively small area, the collector, who assembles animals from trappers and the exporter who is responsible for the management of exports. Smooth functioning of this entire operation depends on close liaison between all members of this chain.

THE TRAPPING AND EXPORT OF MACAQUES IN INDONESIA

C.L. DARSONO

Foundation for the Development of Wild Life Resources, Indonesia.

INTRODUCTION

   The export of substantial numbers of macaques from Indonesia began in 1970 and has continued to increase steadily even though the trapping and transport of these animals is poorly organised. Most animal dealers in Indonesia are not interested in the fate of the animals they sell. The destination of the animals, and the purpose for which they are purchased, are immaterial to dealers, as their sole concern is to make money. They have no idea where their merchandise came from or how it was obtained. Animals are bought from suppliers who contact agents who in turn go from village to village buying animals from farmers.
   Primate exporters in Jakarta do not have access to a continuous supply of primates. They are not aware that they are dealing with an exhaustible resource and that they will eventually have nothing to sell because of their unscientific methods and the destruction of the natural habitats required for the maintenance of wild primate populations.
   Even though the exploitation of primates in Indonesia is poorly organised, a considerable number of *Macaca nemestrina* and *Macaca fascicularis* have been exported (Table I).

TABLE I

Export Statistics of Primates from Indonesia
(Compiled by The Directorate of Nature and Wildlife Conservation in Bogor, Indonesia)

Year	Species	
	M. nemestrina	*M. fascicularis*
1970	1167	6101
1971	464	7650
1972	805	22505
1973	954	21900
1974	1375	12376
1975	1194	15800
Total	5959	86332

   If a long-term supply of this exhaustible resource is to be maintained, then primate procurement must be properly managed and have the support of the importing countries. As a result of the diminished number of primates being exported from India and South America, a greater demand for Indonesian primates is to be expected. Hopefully this will stimulate the establishment of breeding colonies in Indonesia and the correct management of wild populations. If such action is not taken soon, Indonesia will lose a valuable resource and those countries requiring macaques for biomedical purposes will lose a valuable source of research animals. The present human population in Indonesia is 130 million and over the next 25 years is expected to increase to 200 million. The demand for more farmland is ever increasing and is destroying primate habitats in Sumatra and Kalimantan. In order to minimise the effect of such destruction forested areas must be preserved and managed wisely. Unfortunately this is not possible at the present time in Indonesia primarily because of a lack of expertise and funds.

COMMON METHODS OF PRIMATE PROCUREMENT

As an example of procurement methods a typical animal requirement can be followed through the various lines of supply. An animal dealer in Jakarta receives an order for 100 female *M. fascicularis* between 3 and 4 kg body weight. He contacts several suppliers in Jakarta who in turn approach their agents in West Java or South Sumatra. The agents send their men out to the villages where there are still monkeys in the surrounding forest and inform farmers they will buy female monkeys. The farmers trap monkeys whether they can be sold or not because they raid their crops and are a constant pest but in this case they can sell the females. They trap the monkeys one at a time using a simple device which is baited with corn or papaya. When a farmer has time he checks his trap each day but sometimes he forgets or it is inconvenient for him to visit it. A trapped animal dies quickly of dehydration or starvation. When the farmer finds he has caught one he transfers any female to a transport box. Males are clubbed to death and may be used to supplement his diet since he cannot sell them. The farmer has to wait until he has 6 to 10 females before it is worthwhile to take time off from his farm work and bring the animals 5 to 10 km out to the road either by ox cart or carrying the transport box with the help of some friends. While waiting for more animals to be trapped, those already collected are given banana, cassava, corn or rice but usually hardly any water. Some animals are so terrified they will not eat or drink. Once the 6-10 monkeys reach a road, the farmer must wait for one of the agent's men to collect the animals which may be several days. The agent puts them into a group cage with others he has collected. Females which are too big, too small, or very sick, are sold to the meat market.

Once the agent collects 50 females of the appropriate weight range he will ship them to the exporter in Java by boat in small wooden transport boxes. A truck is chartered for the final leg of the trip to Jakarta where the animals are sold to an animal dealer and placed in holding compounds until the appropriate export permits and licences are acquired. The group cages or holding compounds leave a great deal to be desired. Animals from different groups and areas are crowded into a small area and fighting is frequent. Only the strongest get adequate amounts of food and the weakest obtain very little or none. Sick and injured animals are sent to the meat market at this stage. Just before departure the animals are transferred to individual transport cages which contain fruit but no water.

And how many monkey lives did the 100 females cost? If we assume the farmer catches as many males as females, 50% are killed. The loss of female lives from trap-site to the agent's compound is 20-25% and at the supplier's compound another 10-15% die. The animals are usually fed better and kept in cleaner facilities in Jakarta but still another 5-8% are lost while waiting for export documents and banking papers which usually takes one week to 10 days.

Now, if we compute the number of animals that were trapped to obtain 100 females we find that 60% to 75% of the animals trapped died. That is 300-375 macaques were trapped to obtain 100 females; when primate importers stipulate a weight range, usually 3-4 kg, then as many as 500 animals must be trapped to fill the order. This is clearly wasteful and the primate users should share the responsibility.

METHODS USED BY PROFESSIONAL TRAPPERS TO OBTAIN MACAQUES

Professional trappers use group traps. These are constructed of wooden poles cut from the forest and are generally 2.5 m x 2.5 m x 2.0 m in size with a hole in the top. A chute less than 0.5 m long leads from the top downward. Corn, banana or papaya is hung on the outside of the trap and also placed inside. If the dominant male enters first the whole group usually follows but if other animals enter the trap first and become frightened, the rest of the group will not enter. Group trapping is more costly than trapping individuals since the trap is more expensive to build and can only be used once or at the most twice. There are several advantages, however, as more animals can be caught at once and transported to the agent's compounds more quickly as compared with the slower method of catching individual animals to make up the required numbers. Also some behavioural and genetic researchers would prefer to work with a 'natural group'. In order to retain the identity of such a group the agent must keep the animals in marked individual cages or have a large pen for each group. The exporter in Jakarta must have similar facilities. This requires a great deal of capital investment by the supplier and exporter and results in more expensive animals.

CONCLUSIONS AND RECOMMENDATIONS

Indonesia's primate resources must be properly managed and not haphazardly exploited. Long range planning and suitable facilities are required to ensure the availability of primates in the future. This requires co-operation between the exporting and importing countries. The management of wild populations through the establishment of breeding colonies is one possible solution to the various problems.

SUMMARY

The export of macaques from Indonesia for zoos, pets and research has increased steadily since 1970 in spite of poor organisation. Dealers who are concerned with neither the welfare of the animals nor their destination are supplied by farmers, who trap what they consider to be agricultural pests, and by professional trappers. The farmers retain only the required animals (killing all the unqualified animals) and their ignorance of correct diet and maintenance together with long delays while they collect financially viable numbers, leads to very high mortality. The professionals trap family groups but also know little about the needs of their captives. Once the macaques reach the dealers, ignorance, inefficient administration and unnecessary delays lead to further losses. Overall the loss rate is 60-70%. The two species involved, *Macaca nemestrina* and *Macaca fascicularis*, are under constant pressure from the rapidly increasing human population. This, plus inefficient and wasteful trapping and care of captives, is threatening the supply in the long run. The management of wild populations through the fostering of breeding colonies is one possible solution to the problem.

RHESUS MONKEY SUPPLY AND SUPPLY LINES FROM
TRAPPER TO USER (QUARANTINE FACILITY)

D.A. VALERIO

*Hazleton Laboratories America, Inc.,
Vienna, Virginia 22066 USA.*

INTRODUCTION

During a visit to India in November 1975, I was able to tour trapping areas in the Pilibhit Forest and visit exporters' facilities in New Delhi. In this paper I will review my impressions and observations of rhesus monkey supply and supply lines, starting with the trapping areas, following monkeys through the Indian exporter to the American importer and finally to the research and testing laboratory. I have only visited India once and even though I have seen slides from various people who have also been there, I do not pretend to be an expert in this area. Having been involved in using and breeding monkeys for many years, I did have some preconceived opinions and visual impressions as to how things were done in India. Some were confirmed during my visit while others proved to be incorrect.

TRAPPING

In the past, monkey trappers in India were independent agents. Today, it appears some trappers are organised by a company and work for them on a full time basis. It is their occupation, therefore, rather than a sideline, as was often the case previously. Each exporting company has a representative who makes arrangements with the trapping firm regarding the number, weight, sex and age of the animals required as well as payment to the trappers. Dr. Wendell Niemann and I observed trapping operations in two different areas, one near the city of Chandigarh (W.N.) and the other in the Pilibhit Forest (D.V.). Four different methods of trapping were observed.

The first method used a circular, camouflaged, anchored net with four poles, guided and rigged to enclose the animal when the trip wire was closed. In the second, the animals were gradually driven or baited to a selected site, usually in low bush country, where a four to five foot high perimeter net fence was erected. A group of men beat the bushes until the monkeys rushed to the perimeter where they were caught by hand as they became entangled in the net. A rectangular box of bamboo of variable size was constructed for the third method and is favoured in villages, airports, and other areas where nets would be difficult to use. The fourth method is employed in the forest. Animals are manoeuvred into two adjacent trees, ten to fifteen feet apart. All except one of the limbs on one tree are removed and a rope or cord is tied to this limb. Animals are then encouraged to leave the first tree by the only escape route possible, the single limb. They are shaken from this limb and fall 50 to 75 feet to the ground, apparently without injury. The monkeys land within an area previously surrounded by a net and are then caught by the trappers. The method may be modified to include a brush pile as a cushion and in some cases a net is elevated from the ground, serving as a 'trampoline' to break the fall of the animals.

At the site of capture, animals are placed in bamboo crates which contain food and water. Many of these crates have partitions so that animals are separated from one another. This is particularly true of breeding stock although it is not yet the rule to have all animals treated in this way. Eliminating direct contact between individuals provides an excellent method of reducing spread of disease.

Animals are taken by truck or jeep from the trapping areas to the nearest railway station. Generally, they arrive in Delhi the following morning and, on arrival, a veterinarian from the export company is on hand to check the shipping conditions of arrivals, e.g. food and water supply, injury, or illness. He then supervises the transport of the animals by van to the export facility, their uncrating and their placement in cages.

EXPORTERS' FACILITIES

T.E. Patterson India (Pvt Ltd) and Vita Private Ltd (VPL) are the two main exporters of rhesus monkeys from India and account for almost all the rhesus monkey export trade. They have affiliations in the UK, the USA, Japan and Europe, giving some continuity to the international movement of animals to the European continent, America and Russia. Both export facilities are near the periphery of Delhi Airport, thus travel time to meet overseas flights is a matter of minutes. These companies have been exporting primates for biomedical research to the various user countries for the past 30 years. Presently, they are changing from intermediaries between the trapper and importer to a role including animal health monitoring and disease control.

Both companies were visited, Vita at two locations including their older facility and their new unit which they have occupied since the early part of 1976. All compounds are very modern and have perimeter walls to keep out intruders from local villages.

The conditions under which animals are held, including veterinary care, overall sanitation, water treatment, provision for isolation and treatment of the injured and sick, as well as tuberculosis testing procedures and record keeping, were judged to be as good as, or better than, most research facilities in India and some in the United States. I, as well as my colleagues who visited these facilities, had a very good impression of the care and surveillance the animals were given, contrary to our opinions prior to visiting India. A health programme, including tuberculosis surveillance of the employees, is either anticipated or in effect at both compounds. In order to meet the requirement of the 30 day quarantine which was imposed by the Indian Government in 1975, both companies are either renovating their present facilities or constructing new facilities. There is a daily census of over 1,000 animals at each of the exporters' premises. Now all animals are being individually caged making it possible to maintain the identity of an animal from time of capture to arrival at the laboratory.

Upon receipt, animals are weighed and placed in rooms according to weight. All are tuberculin tested and those requiring medical treatment are separated and placed in an intensive care room. At one facility, animals are fed a commercial pellet diet and at the other a combination of various native grains, fruit and vegetables. Nutritionally, both regimens appeared more than adequate.

IMPORTERS' FACILITIES

In the United States there are two main importers; Primate Imports, a subsidiary of Charles River, and PrimeLabs, which recently became a subsidiary of Hazleton Laboratories Corporation. Both facilities are located in the northeast near the New York airports and import not only rhesus monkeys but also other non-human primates from Southeast Asia, including long-tail, pigtail and stump-tail macaques. They also import various species from Africa, and New World monkeys, when available, from South America.

There is no regulation in the United States determining how long an importer holds animals before shipment to the research laboratory. This is entirely dependent upon the client's demands. The large primate users in the USA prefer to take animals direct since they feel they have comparable or better facilities than the importer and more time to spend in treatment and conditioning of the animals. Arrangements are made for these animals by the importer, but when they arrive in New York, they usually go directly to the user's receiving facility, although some animals may stay at the importer's premises for a few days before being sent to the user. Other institutions prefer the 'conditioned monkey', that is one which has stayed at the importer's facility 30 to 45 days. The conditioned animal is then sent to the user where it is placed in their quarantine facility for a specified period of time.

Both Primate Imports and PrimeLabs have been visited many times over the past several years. The following is my understanding of what procedures are followed at these facilities.

Animals are received after shipment from Delhi, usually via London to a New York airport. They are taken to the importer's facilities in air conditioned or heated vans, depending on the time of year. Upon receipt, they are weighed and placed in individual cages in one or more rooms, depending on the number of monkeys received and their weights. The animals are generally tuberculin tested immediately, or within a few days of arrival. They are examined and any animal that appears unhealthy or has injuries is referred to the resident veterinarian. Animals are tuberculin tested every two weeks when they are weighed. Positive reactors are killed. An anthelmenthic is administered by intranasal gastric intubation which is repeated two weeks later if it is to be a conditioned animal. Screening procedures involving haematology, biochemistry and x-ray tests would be undertaken on a specific request from the client. Complete records are maintained on all animals.

RESEARCH AND TESTING LABORATORY QUARANTINE FACILITIES

The basic principles of design and construction of a primate quarantine facility need not differ significantly from those usually considered desirable for other primate facilities or for other species of animals. There are, however, several details in the design and construction which are important and, if ignored, can increase the danger to the health of both personnel and animals, make for difficulties in the day-to-day management and contribute significantly to operating costs.

Most users' quarantine facilities are separate buildings and not an integral part of the laboratory building. There are, however, some large establishments in the USA which have their quarantine facility within the research or testing building. It will be at one end of the building, and is generally separated from other animals and laboratories by at least two corridors. At many institutions, such as universities and small government facilities, the need for primates for biomedical research is so small that only one or two rooms are set aside for a primate quarantine facility. Therefore, location is mainly dictated by the number of animals used and the type of research or testing that is performed.

One of the principal hazards of maintaining a primate quarantine unit is the problem associated with certain bacterial and viral diseases, and occasionally mycotic and parasitic infestations. Biohazards are particularly common with primates recently imported from their natural habitat. Such organisms include *Mycobacterium tuberculosis*, *Salmonella* and *Shigella* species, herpes B, and Marburg virus agents. It is also possible that humans infected with one or other of the above organisms could transfer them to the animals. Because of these factors, it is necessary to design a unit for monkeys in such a way that the risk of transmitting infections among themselves, to personnel, and vice versa is reduced to a minimum. For details on a primate quarantine facility design, including recommended air handling systems, see the recent article by Valerio (1975).

Quarantine procedures that are generally used at the user facilities in the United States are as follows. Cages are generally cleaned by a mechanical cage washer. Due to the importance of tuberculosis, most facilities use phenolic fog in all rooms once they are emptied of a quarantine shipment. Most users have several quarantine rooms of various sizes, but generally the largest holding capacity is for 75 monkeys. Fifty monkeys per room or less is perhaps the ideal room size for the quarantine of animals.

Upon receipt animals are taken from their crates and are usually given a physical examination by an attending veterinarian for any signs of illness or injury, and treated on an individual basis. The animals are tuberculin tested, weighed and individual health records are initiated. The animals are also assigned a number, which is tattooed on their chests using a consecutive numbering system. Generally, all of this is done on the first day. Some people prefer to wait, however, until the animal is acclimatised to its cage and environment. At the National Institutes of Health's quarantine facility in Poolsville, the animals are put directly into cages on the first day and nothing else is done at that time.

Animals are fed a commercial, 25% protein monkey chow, which is usually supplemented with fruit several times a week during the quarantine period until they are acclimatised. Some institutions use a 15% protein diet. I personally feel there is little difference. If the animals are to use automatic waterers, one must be sure that they know how to get water from the valves. Often the animals are dehydrated upon arrival and they can easily get worse in a very short time unless this is observed. Sometimes one must hold a rod or stick against the valve so that the animal can see where the water comes from.

At the second and fourth week tuberculin tests, the animals are given an anthelmenthic (thiabendazole, 100 mg/kg) by pediatric stomach tube. A prophylactic IM injection of benzathene and procaine-penicillin (300,000 international units) is given twice. If older animals (breeders) are being received, they get a chest radiograph upon entry, or at least within the first two weeks in the quarantine unit. This is done mainly to eliminate any potential anergic reactor. A chest radiograph is obtained on all animals between 70 and 90 days in the quarantine phase. Tuberculin tests are continued at two week intervals throughout the entire quarantine period and a minimum of five consecutive negative tests are required before a group of animals is released for research. Animals are examined daily and records are maintained for any clinical signs of disease - anorexia, nasal discharge, diarrhoea, dysentry, etc. Animals are treated accordingly depending upon the diagnosis. After two administrations of an anthelminthic, random faecal samples are checked for the effectiveness of the anthelminthic on internal parasites. If needed, treatments are repeated.

The length of our quarantine period for all monkeys is a minimum of 90 days. This length of time is mainly to minimize positive tuberculosis cases from leaving the quarantine facility and entering the user's research facility. Many people have used shorter periods but eventually have had disaster strike once a research or testing programme is underway.

RHESUS MONKEY SUPPLY: QUOTAS

There was divergent opinion among the participants at the symposium entitled 'The Use of Non-human Primates in Biomedical Research' which was sponsored by the Indian National Science Academy in November 1975, concerning the number of rhesus monkeys available. No better consensus of opinion exists some nine months later as to what can be expected regarding the Indian quota or the number of rhesus monkeys in India. At the symposium opinions varied from the need for a total export ban for five years to increasing the present quota of 20,000 animals per annum by a factor of 2.5 (50,000). In general, there has been no reliable animal census from which to compute a reasonable quota from the wild without causing irrevocable harm to the rhesus monkey. This dire need was well recognised during the symposium and, therefore, field studies and a population census were recommended as essential in order to have a comprehensive view of the wild population in support of an annual quota.

The expansion of agricultural land and the gradually increasing intolerance of the monkey population by a younger generation of people are additional factors which should be taken into account but are difficult to define in terms of numbers of animals.

People who have visited India within the last ten years have claimed that endless numbers of monkeys could be seen along roadsides. In my limited travel of approximately 500 miles within a 150 mile radius of Delhi, however, I saw less than 50 monkeys along the road in five days of travel. Thus, the roadside groups along the major highways we travelled were not evident. The trapping sites we visited were a 1.5 to 2 mile walk off secondary roads, indicating to me that the more readily accessible groups have disappeared.

One can therefore conclude that concern about reduced numbers is a very real one. Among those who went to the various trapping areas, our single opinion is that it would be extremely difficult to estimate the number of monkeys that reside in the forest areas of India. Therefore, the annual export quota now set at 20,000 may be arbitrary, but at present, there are insufficient data to support any other figure.

The concept of a controlled trapping programme, perhaps along the lines of the successful wildlife management programmes, was described by Southwick (1977) as a viable alternative. One positive aspect of this suggestion is that because the forest monkeys, in contrast to the village or roadside groups of animals, have a limited contact with man and domestic animals (primarily oxen), a generally healthier animal should be obtained. In addition the exporters are helping to improve the quality of supplies by currently investing large amounts of capital in order to upgrade their facilities. The effect of the 30 day quarantine in India on newly imported monkeys remains to be seen. Certainly some diseases, e.g. measles and acute pneumonia, will now probably be resolved at the exporter's premises. On the other hand, 30 days is insufficient time to prepare an animal for shipment overseas and, at that point, further stress may be deleterious to the animal's health. The Indian Government is presently studying the need for a 'Certificate of Movement' for the trapper, probably to be issued by the Chief Wildlife Warden. Depending on how this is to be operated, delays following trapping could occur while certificates are being issued. In this case, the quality of the animals could be reduced due to inadequate holding capabilities.

IMPROVEMENTS TO BE MADE AND OTHER SPECULATIONS

One must always reflect on the total system when considering which elements in the supply line could be improved. It is known, based on what has been covered at this meeting and what we have heard in the past, that the systems of trapping rhesus monkeys, as well as other species, have definitely been improved over the years. Recently, I was extremely impressed with the various systems. Yet, I am still concerned about the amount of tuberculosis that is seen upon receipt of animals in the user's facility. Regardless of species, one must, through an education process, try to convince trappers and trapping organisations that not every animal captured should be passed on along the supply line as sometimes an animal is not the correct age, is diseased, or has an injury or abnormality. This remains a problem since trappers are often paid on a per capita basis, hence although an animal may be worthless from a research standpoint, it is passed along the supply line.

Once again, the export facilities I visited were extremely modern and are difficult to criticise. The only areas of improvement that I can suggest are (i) animal groups should be separated and (ii) tuberculin antigen should be consistent. I raise the following questions: (a) are proper procedures being carried out respecting tuberculosis eradication, i.e. correct and proper frequency of tuberculin administration; (b) is a reliable antigen being used comparable with that available in the USA and Europe?

Another problem is that although the importer usually receives what was ordered (sex, age, condition) from the exporter, there always seems to be undesirable animals (e.g. under-weight, old, poor condition, diseased) included. Again, this is a case where the exporter has paid the trapper on a per capita basis for animals. The exporter, not wanting to take the loss, passes the animal to the importer, who then faces the same problem. Whether these losses should be spread along the entire chain of supply, or should be absorbed by the trapper is difficult to ascertain. However, I am only raising some issues which I believe are inherent in the present system, and should be given attention in the future.

At the importing facilities, there are areas where the need for high quality procedures should be stressed, e.g. personnel procedures, good air handling systems, and proper room separation to minimize diseases. Procedures must be conducted properly and regularly, primarily in relation to tuberculosis and diseased or unhealthy animals should not be passed on to the user.

Certainly there is a need for complete communication throughout the line of supply. The scientist and/or research facility, must pass information on to the importer who, in turn, must pass it on to the exporter, and the exporter then has to get this information to the trapper. If each step is not taken, one particular element can be making the same mistake, unintentionally, simply by not being informed. Improvements in the supply line cannot be expected unless there is good communication. This is particularly important since there are so many elements involved in the route of the monkey from the forest to the research facility.

SUMMARY

Indian monkey exporting companies now provide a sophisticated service, employing full-time trappers and ensuring that high standards of disease control are maintained. After arrival in the USA animals may go directly to the research institute or undergo conditioning at the importer's premises prior to delivery to the user. Conditioning includes screening for tuberculosis and elimination of internal parasites. At the research unit, care must be taken to ensure that cross-contamination is prevented by application of modern building construction methods, good management and sensible technology. There is need for a census of the rhesus monkey population in India, thereby allowing a rational export quota to be decided.

ACKNOWLEDGEMENT

I would like to thank Dr. Wendell Niemann for contributing some of the information on trapping contained in this manuscript.

REFERENCES

Southwick, C.H. (1977). *In* "The Use of Non-human Primates in Biomedical Research" (M.N. Prasad and T.C. Anand Kumar, eds). University of Delhi, in press.
Valerio, D.A. (1975). *In* "Proceedings of the National Cancer Institute Symposium on Biohazards and Zoonotic Problems of Primate Procurement, Quarantine and Research" (M.L. Simmons, ed.), pp. 67-78.

INTERNATIONAL TRAFFIC IN PRIMATES FROM THAILAND

ARDITH A. EUDEY

*Department of Anthropology, University of California,
Davis, California 95616 USA.*

INTRODUCTION

The biomedical establishments in the United States have been the major recipients of primates exported from Thailand, and during the period 1964-1973, for which figures are available, Thailand ranged between fourth and tenth as a supplier of primates to the United States (Committee on Conservation of Non-Human Primates, 1975). Macaques (*Macaca* spp.) have been the principal export.

In 1972 the Thai Department of Foreign Trade began to record the export of macaques by species (Table I). Previously records were maintained only for the genus *Macaca*. Many 'unclassified' primates were included in the figures compiled for 1972-1973, but by 1974 the problem of specific identification appeared to be largely overcome.

PUBLISHED DATA FOR EXPORTS FROM THAILAND

In 1974 a total of 5,291 macaques were exported from Thailand (Table I). Of this total, 3,904 (74%) were exported to the United States. Stumptail macaques (*Macaca arctoides*) accounted for 2,398 (45%) of the exports, with 1,816 (76%) being exported to the United States, the residue going to five other countries.

Unpublished figures compiled from 1974 import declarations for the US Department of the Interior by A.M. Greenhall, record the importation from Thailand of only 475 macaques, of which 175 were stumptail macaques. He attributed the disparity to the fact that some declarations never reached his office. No figures are yet available for 1975.

In 1975 a total of 2,998 macaques was exported from Thailand (Table I). An additional 90 animals were designated 'unclassified' and may include the slow loris, *Nycticebus coucang*. Of the known macaques, 2,020 (67%) were exported to the USA. Stumptail macaques accounted for 1,675 (56%) of the exports, with 1,042 (62%) of these exported to the United States. During 1975 eight other countries received shipments of stumptail macaques. Although the total number of macaques exported from Thailand decreased in 1975 by over 2,000 from that exported in 1974 (and previous years), increases occurred in both the percentage of stumptail macaques exported and the number of countries importing the species.

In the United States the stumptail macaque may be the fifth most commonly used primate in biomedical research (Committee on Conservation of Non-Human Primates, 1975) and almost all animals are obtained from Thailand (Clapp, 1974). A bibliography on laboratory studies of *Macaca*

TABLE I
Macaque Exports from Thailand – 1975[a]

Country	Rhesus (M. mulatta)	Pigtail (M. nemestrina)	Crab-eating (M. fascicularis)	Stumptail (M. arctoides)	Unclassified[b]	Total
USA	235	585	158	1042	60	2080
Japan	30		95	211		336
France				20		20
Italy	40	25	20	30		115
Holland		10				10
England	40	10	30	180	30	240
Austria				5		5
Greece			10			10
Bolivia				5		5
West Germany			30	180		210
Lebanon		2	3	2		7
Total	345	632	346	1675	90	3088

Macaque Exports from Thailand – 1974[a]

Country	Rhesus	Pigtail	Crab-eating	Stumptail	Unclassified	Total
USA	1111	585	392	1816		3904
Japan	150	133	100	212		595
France	40	24	180	160		404
Italy	115	35	1	45		196
Holland		6		80		86
West Germany			10	85		95
Czechoslovakia	11					11
Total	1427	783	683	2398		5291

Macaque Exports from Thailand – 1973[a]

Country	Rhesus	Pigtail	Crab-eating	Stumptail	Unclassified	Total
USA	217	167	235	641	2077	3337
Japan	34	42	110	20	322	528
France	20		100	20	201	341
Italy	40	44	18	66	147	315
Holland			10		40	50
West Germany			10		30	40
Belgium					160	160
Switzerland			1			1
Hungary					10	10
India			1			1
Total	311	253	485	747	2987	4783

Macaque Exports from Thailand – 1972[a]

Country	Rhesus	Pigtail	Crab-eating	Stumptail	Unclassified	Total
USA	6		1	318	4858	5183
Japan					129	129
France			6		25	31
Italy		10	24	10	64	108
Holland					1	1
Greece					4	4
Total	6	10	31	328	5081	5456

[a] Figures compiled by the Department of Foreign Trade (Export-Import Division), Ministry of Commerce, Royal Thai Government.
[b] This classification may include the slow loris as well as macaques.

TABLE II

US Institutions using Stumptail macaques, 1971-1976, as derived from a Bibliography on Laboratory Studies of Macaca arctoides *compiled by the Primate Information Center, Regional Primate Research Center, University of Washington.*

Kind of Facility[a]	Number of Institutions	Combined Number of Published Studies
Regional primate research centers	7	51
National Institutes of Health (and affiliated institutions)	5	18
US military	4	12
State government	1	3
Private research foundations	6	12
Hospitals	9	26
Medical schools	40	115
Professional schools	9	12
Veterinary schools	4	5
Laboratory animal programs, medical schools	2	3
Universities	22	62
Industry	3	4
Total	113	323

[a]Some differentiation of facilities is arbitrary due to the existence of separate research units at some universities and the multiple use of primates at others.

arctoides for 1971-1976 compiled by the Primate Information Center, Washington Regional Primate Research Center, establishes the use of stumptail macaques in at least 113 separate facilities in the United States and in at least 20 countries (Tables II and III).

All available evidence indicates that the stumptail macaque is not plentiful enough to withstand continued exploitation such as that represented by the annual importation into the United States alone of over 1,000 animals (Wolfheim, in preparation). In 1970, Harrisson (1971) had already identified *Macaca arctoides* as an inadequately studied and potentially endangered species. Throughout much of its distribution, especially in Indochina, the species has experienced increased hunting pressure and habitat destruction, including defoliation, as a consequence of prolonged warfare. Harrisson (1971) implored 'non-Asian researchers to make every effort to originate or support ecological studies on stumptail macaques and, simultaneously, reduce the use of the species and promote laboratory breeding'.

In spite of this exhortation, the number of stumptail macaques maintained in US laboratories increased from approximately 809 to 1,083 in 1973, of which only 287 had been maintained for 3 or more years (Committee on Conservation of Non-Human Primates, 1975). On 1st October 1973, only 260 adult females were maintained in breeding colonies in the United States and only 68 births had been recorded for the year (Committee on Conservation of Non-Human Primates, 1975).

TABLE III

Countries (other than the US) using Stumptail macaques, 1971-1976, as derived from a Bibliography on Laboratory Studies of Macaca arctoides *compiled by the Primate Information Center, Regional Primate Research Center, University of Washington.*

Country	Number of Institutions	Combined Number of Published Studies
Australia	2	2
Canada	4	5
Denmark	1	1
England	8	11
Finland	1	2
France	8	10
East Germany	1	1
West Germany	4	4
India	1	1
Italy	1	3
Japan	4	10
Netherlands	5	10
Norway	1	1
Poland	1	1
Spain	1	1
Sweden	1	3
Switzerland	4	7
Thailand	1	1
USSR	3	16
Total	52	90

A total ban on the export of primates from Thailand became effective in April 1976. Khun Pong Leng-ee (personal communication), Chief of Wildlife Conservation, reported that animal dealers in Thailand are protesting about the ban in anticipation of financial loss from the termination of sales of some macaques, especially *Macaca arctoides*.

In 1976 the US Fish and Wildlife Service proposed that *Macaca arctoides* be designated a threatened species. The need for protection in both exporting and importing countries is well illustrated by the illegal traffic in gibbons, especially *Hylobates lar*, which existed between Thailand and the United States. In 1965 the Thai government banned the commercial exploitation and export of all gibbons. The 1974 commercial acquisitions of *H. lar* from Thailand by the Comparative Oncology Laboratory, School of Veterinary Medicine, University of California, Davis, are well documented, however, and currently under investigation by the US Fish and Wildlife Service. The Comparative Oncology Laboratory also received a shipment of gibbons from the US Army in Thailand in 1973 under unusual circumstances. Strong circumstantial evidence suggests that other US laboratories as well as commercial animal dealers may have received illegal commercial and/or military shipments of gibbons from Thailand subsequent to 1965.

All gibbon species appear on Appendix I of the Convention on International Trade in Endangered Species. On 14 June 1976 the United States recognised all gibbons as endangered, but it now appears that the National Institutes of

Health is attempting to circumvent both Thai and US laws by supporting a conservation and breeding program in Thailand to obtain gibbons for export to the United States.

SUMMARY

The United States has been the major recipient of primates exported from Thailand. Macaques, especially *Macaca arctoides*, were the principal export until a total ban on the export of primates became effective in April 1976. Although *M. arctoides* was recognised as potentially endangered in 1970, it may be the fifth most commonly used primate in US biomedical research, and most of the animals were obtained from Thailand. In 1976 the US Fish and Wildlife Service proposed that *M. arctoides* be designated a threatened species. The need for protection in both exporting and importing countries is illustrated by the illegal traffic in gibbons which has existed between Thailand and the United States.

ACKNOWLEDGEMENTS

I wish to thank the National Research Council of Thailand, the Wildlife Conservation Division, Royal Thai Forest Department, and the Primate Information Center, Regional Primate Research Center, University of Washington, for assistance.

REFERENCES

Clapp, R.B. (1974). Mammals imported into the United States in 1972. Special Scientific Report - Wildlife No. 181. US Government Printing Office, Washington.
Committee on Conservation of Non-Human Primates (1975). "Non-human Primates; Usage and Availability for Biomedical Programs". Institute of Laboratory Animal Resources, National Academy of Sciences, Washington.
Harrisson, B. (1971). "Conservation of Non-human Primates in 1970". S. Karger, Basel.
Wolfheim, J.H. (in preparation). "The Status of Wild Primates". US Fish and Wildlife Service, US Government Printing Office, Washington.

TRANSPORT OF PRIMATES BY AIR

G.E. JOSS

8 Whitepost Hill, Redhill, Surrey UK.

INTRODUCTION

Most members of this Congress will have had some experience in transporting primates from one part of the world to another by air. It may be of interest for you to hear how the present regulations relating to air travel for these animals came into being.

During the early 1950's there was a sudden huge demand for rhesus monkeys from India for poliomyelitis vaccine production. Although importers made valiant efforts to ensure that these animals were carried humanely, lack of prior experience and knowledge resulted in an unacceptably high death rate of about 15%. Deaths were caused by one or other of the following factors:
1. Monkeys were sick or emaciated prior to despatch from India.
2. Handling was often rough and prolonged.
3. Conditions of housing and hygiene at the exporters' farm were poor, leading to dissemination of disease-producing organisms.
4. Travelling boxes were badly designed, which resulted in fighting and inadequate ventilation. Excessive numbers were often packed in these travelling boxes.
5. Animals were often subjected to extremes of temperature during transit at airports or at refuelling stops.
6. Insufficient attention was paid to ventilation of baggage holds when aircraft were parked on the ground.
7. Transit stress, spread of disease and losses were frequently exacerbated by delays at airports.

INTERVENTION BY BRITISH STANDARDS INSTITUTION

Because of adverse publicity, airlines sought advice from persons who were thought to have the required knowledge, but it was soon apparent that advice was being given on a subject about which very little was known. Therefore it became a matter of urgency to investigate the matter with a view to introducing guide lines for both airlines and those persons involved in the trade.

Accordingly, and greatly to its credit, the British Standards Institution (BSI) formed a committee of interested parties and this body produced the British Standard BS 3149 (1959) entitled 'The Carriage of Monkeys by Air'. This is now an historic document but it was then a valuable guide to the airlines carrying monkeys from India. Subsequently the Indian Standards Institution produced a similar document and the provisions outlined became mandatory for airlines shipping monkeys out of India. As a consequence of enforcement of these standards, transit deaths were reduced from 15% to less than 1%.

EFFORTS TOWARDS IMPROVEMENTS IN TRANSIT CONDITIONS

Monkeys arriving from Delhi are nowadays invariably in good condition but this cannot be said for animals from other parts of the world. In 1966 the International Air Transport Association (IATA) formed a study group to examine the whole question of the air transport of animals generally. It consisted of airline representatives from many countries, with myself as veterinary adviser. Two years later, and following much discussion and collecting of information, the first IATA Live Animals Manual was produced with recommended minimum standards for containers for over 200 different species, and suggestions for handling and care of these animals in their containers. The fourth edition of this document, published in February 1975, was recognised by all IATA airlines and the recommendations outlined were accepted as mandatory. Some information and instructions set out in the IATA manual may seem elementary, but it must be remembered that it is written for the information of airline personnel of all nationalities.

Enforcement of the IATA regulations has proved to be both very difficult and unsatisfactory and has moreover led to much ill-feeling. This has been in part due to a lack of interest and co-operation by the governments concerned and the casual attitude of some airlines. Another unsatisfactory feature of the system has been the handling agents that airlines are obliged to employ, because they are appointed by local government and are difficult to regulate.

British airlines have made special efforts to obtain full co-operation from foreign governments but, despite personal approaches, it has not been possible to achieve a full understanding. In June 1976 as a result of representations made by veterinary staff of the Ministry of Agriculture in London, a meeting of heads of government veterinary departments was held in Geneva. It was agreed at this meeting that delegates would take back information to their respective governments concerning the air transport of animals and insist that their own national airline observe the IATA regulations. It remains to be seen what effect these representations will have.

RECOMMENDATIONS TO IMPROVE AIR TRANSPORT OF PRIMATES

In order to ensure that primates are carried from place to place by air both safely and humanely, IATA airlines now insist that the shipper signs a certificate giving the species and number of animals in the batch, that they are in good health, fit to travel and provided with a container consistent with the recommendations in the IATA manual. It is further suggested that:
1. Bookings should be made well in advance.
2. The consignee should be informed of date and time of arrival of the animals.
3. Sick or injured animals should not be despatched.
4. Animals should be met at airports and reliance not placed on hostels or cargo centres.
5. Arrangements for customs clearance should be made in advance.
6. Cargo officers at airports should ensure that air supplies to animals in cargo holds are adequate.

SUMMARY

Twenty years ago the carriage of monkeys by air was grossly unsatisfactory.

Initiatives by the British Standards Institution, Government Departments and the International Air Transport Association led to great improvements in standards not only for simian primates but for all species of animals carried by air.

REFERENCES

British Standards Institution. (1959). Carriage of Live Animals by Air: Monkeys for Laboratory Use. BS 3149 Part I.
International Air Transport Association (IATA). (1975). Live Animals Regulations. Aerad Printers.

PRIMATE IMPORTS INTO THE UNITED KINGDOM 1965-1975

J.A. BURTON

Fauna Preservation Society, c/o Zoological Society of London, Regent's Park, London NW1 UK.

INTRODUCTION

In 1964 the Restriction of Import (Animals) Act (1964) came into force and by 1965 was operational. This legislation was designed to control imports of rare (live) mammals and reptiles. It remained in operation until the end of 1975 when it was suspended. Since that date some controls have been in force under the Special Powers Act (1939), wartime legislation which can be used to control any imports and it was under this legislation that the United Kingdom ratified the Washington Convention. By the end of 1976, consolidating legislation had been passed and, as far as primates are concerned, the United Kingdom complies fully with the requirements of the Washington Convention (1973).

Under the 1964 Act animals were divided into two lists, very like the lists of the Washington Convention, both in their content and in the way they were administered. This note is based almost exclusively on the annual reports of the Advisory Committee established under the 1964 Act.

It is my belief that in general terms the various restrictions have been reasonably well enforced. There has undoubtedly been some evasion because cases of smuggling have come to light. I do not believe, however, that this alters the general picture significantly in spite of the fact that the illicit animals are often of the rarer species. I believe that at present the most serious threats posed by United Kingdom primate imports are to the more numerous species.

SPECIES AND NUMBERS IMPORTED

During the 11 years 1965-1975 some 180,000 primates were imported into the United Kingdom. Prior to 1965 there is little information and that given for 1965-1967 throughout this paper refers to licences issued, not actual imports, which exaggerates the totals slightly. In this period six families of primates have been imported, namely, Pongidae, Tarsiidae, Lemuridae, Callitrichidae, Cebidae and Cercopithecidae. The first three were in very small numbers (Table I) and total just over 350 individuals. The other three families form the bulk of the trade and from 1965-75 over 8,500 marmosets (5% of total) nearly 24,000 cebids (13%) and some 147,000 cercopithecids (82%) were imported (Table II). Table III shows how, over the years, the percentage of licences being taken up has decreased from a maximum in 1970-71, to an all-time low in 1975. This might be because supply is failing to meet demand.

The declared purpose of the majority of imports (Table IV) was scientific

TABLE I

Importations into the United Kingdom 1965-1975

	1965[a]	1966[a]	1967[a]	1968	1969	1970	1971	1972	1973	1974	1975	Total
Lemuridae	4	1	1	2	34	6	-	5	16	-	1	70
Tarsiidae	60	10	4	-	-	1	-	-	-	1	-	66
Pongidae	84	52	57	26	59	6	4	31	42	7	7	375

[a]Figures for years 1965-67 inclusive relate to licences issued.

TABLE II

Imports of Callitrichidae, Cebidae and Cercopithecidae into the United Kingdom for 11 years, 1965-1975 Inclusive shown as Numerical Totals and as Percentages

	1965[a]	1966[a]	1967[a]	1968	1969	1970	1971	1972	1973	1974	1975	Total	Average
Callitrichidae	1507 4.8%	2112 7.4%	1278 4.9%	429 3.6%	437 3.6%	413 3.5%	322 2.5%	339 2.6%	1040 9.0%	517 4.1%	163 2.0%	8557 4.8%	778 pa
Cebidae	6444 20.7%	5360 18.8%	3697 14.1%	1543 12.8%	1726 14.2%	892 7.6%	809 6.2%	1155 8.8%	589 5.1%	1321 10.5%	500 6.1%	24035 13.3%	2185 pa
Cercopithecidae	23176 74.4%	21066 73.8%	21198 81.0%	10070 83.6%	9978 82.1%	10376 88.9%	11827 91.3%	11557 88.6%	9960 85.9%	10678 85.3%	7487 91.9%	147373 81.9%	13397 pa

[a]1965-67 inclusive relate to licences issued. The imports were probably somewhere between 50-70% of these figures in reality.

TABLE III

The Percentage of United Kingdom Import Licences taken up, 1968-75 Inclusive

	1968	1969	1970	1971	1972	1973	1974	1975
Callitrichidae	66.3%	54.7%	99.5%	100%	86.4%	43.9%	29.9%	23.0%
Cebidae	55.9%	62.6%	92.9%	86.3%	82.5%	39.7%	37.0%	29.0%
Cercopithecidae	60.7%	60.5%	72.2%	78.0%	69.6%	55.5%	50.0%	48.3%

TABLE IV

Declared purpose of Importation, expressed as Percentage of Total for each Family

Family	Breeding and Exhibition (including Zoos)		Scientific Research (including vaccine production and testing)		Pets (of importers)		Resale (including resale to other headings)	
	1965	1974	1965	1974	1965	1974	1965	1974
Callitrichidae	0.4%	6%	40%	94%	0.6%	0%	59%	0%
Cebidae	4%	6%	46%	86%	70.1%	0%	50%	8%
Cercopithecidae	0.8%	1%	89%	99%	0.2%	0%	10%	> 1%

research which included the production and testing of vaccines. In 1975, 17,899 primates were imported for research, but only 259 for exhibition (which includes zoos and breeding). None were imported for resale or as pets.

It must be borne in mind that stricter controls under disease regulations, notably those to prevent the introduction of rabies (Rabies {Importation of Dogs, Cats and other Mammals} Order, 1974) into the United Kingdom, have also had a considerable effect, particularly on the pet trade. The strict quarantine regulations which apply to the import of all primates have helped to force prices up and there has probably been a consequent fall in demand even for scientific and medical research purposes. Nevertheless, should such restrictions be eased there is little doubt there would be a sudden increase in imports provided that animals were available.

During the period under review there has been a marked shift in the declared purposes for which the animals were imported. Table IV shows the declared purpose of the imports of Callitrichidae, Cebidae and Cercopithecidae. Most noticeable is the fact that by 1974 there were no imports of primates as pets, only 103 Cebids and 1 Cercopithecid were for resale while all others were imported direct by the consumer. In 1975 all were apparently imported direct, although the status of the specialist primate dealers in the United Kingdom is not clear from the reports.

Pongidae

368 individuals of 10 species imported of which *Pan troglodytes* were the most numerous (Table VI). The majority of the imports occurred from 1965-69, and since that time there has been a drastic reduction. Gibbons have from time to time been smuggled but illicit trade into the United Kingdom is very small and fairly easily detected while that passing through London is unlikely to have a significant effect on wild populations at its present level.

Tarsiidae

Apparently none have been imported since 1967 and the totals given are suspect as statistics for 1965-67 refer to licences issued, not actual imports.

Lemuridae

A total of 70 individuals of 8 species were imported of which *Lemur catta* (25) and *Microcebus murinus* (26) were the most numerous. The majority of Lemurids imported since 1969 have been captive bred.

Callitrichidae

8,560 individuals of 11 species (17 subspecies) were imported. I have followed the nomenclature of Napier's catalogue of the Primates in the British Museum (Natural History) part 1 (1976). The nomenclature used in the United Kingdom statistics is somewhat confusing and so I have listed all subspecies in Tables V, VII and VIII.

The most numerous imports were those of the three races of *Callithrix jacchus* totalling 5,877 (69%). In line with the strict conservation measures proposed for many species, United Kingdom imports have dropped recently and in 1975 only four species, namely *Callithrix jacchus auritus*, *Saguinus oe. oedipus*, *S. labiatus* and *Callithrix argentata*, were imported. However, it is

TABLE V
Imports of Cebidae (1965-1975 Inclusive) into the United Kingdom

Alouatta villosa	100
Alouatta seniculus	139
Alouatta belzebul	37
Alouatta caraya	13
Alouatta fusca (including *guariba*)	24
Aotus trivirgatus	1182
Ateles paniscus	883
Ateles belzebuth	112
Ateles fusciceps	101
Ateles geoffroyi	492
Brachyteles arachnoides	24
Cebus capucinus	2051
Cebus albifrons	553
Cebus apella	942
Cebus nigrivittatus	150
Cacajao rubicundus	7
Callicebus personatus (including *melanochir*)	14
Callicebus moloch (including *cupreus*)	11
Lagothrix lagotricha (including *cana*)	1609
Lagothrix flavicauda (including *hendui*)	100
Pithecia monachus	32
Pithecia pithecia	194
Saimiri sciureus	15145
Saimiri oerstedii	120
Total	24035

TABLE VI
Imports of Tarsiidae, Lemuridae and Pongidae (1965-1975 Inclusive) into the United Kingdom

Lemuridae		Pongidae		Tarsiidae	
Lemur fulvus	6	*Gorilla gorilla*	39	*Tarsius syrichta*	10
Lemur catta	25	*Hylobates moloch*	5	*Tarsius bancanus*	50
Lemur macaco	1	*Hylobates lar*	50	*Tarsius spectrum*	4
Lemur mongoz	1	*Hylobates hoolock*	8		
Lemur variegatus	1	*Hylobates agilis*	3		
Cheirogaleus medius	10	*Hylobates syndactylus*	21		
Microcebus murinus	26	*Hylobates concolor*	1		
		Hylobates spp.	13		
		Pan troglodytes	202		
		Pan paniscus	1		
		Pongo pygmaeus	25		
Total	70	Total	368	Total	64

TABLE VII

Imports of Cercopithecidae (1965-1975 Inclusive) into the United Kingdom

Cercocebus albigena	19	*Macaca radiata*	13
Cercocebus torquatus	166	*Macaca cyclopis*	31
Cercocebus aterrimus	3	*Macaca sylvana*	8
Cercopithecus nictitans	269	*Macaca sinica*	5
Cercopithecus cephus	278	*Mandrillus leucophaeus*	144
Cercopithecus neglectus	373	*Mandrillus sphinx*	719
Cercopithecus diana	443	*Nasalis larvatus*	2
Cercopithecus mona	602	*Papio cynocephalus*	4352
Cercopithecus aethiops	11323	*Papio doguera*	2853
Cercopithecus talapoin	600	*Papio papio*	11940
Cercopithecus mitis	109	*Papio ursinus*	1069
Cercopithecus petaurista	5	*Papio hamadryas*	1043
Cercopithecus niger	28	*Papio anubis*	751
Cercopithecus l'hoesti	6	*Presbytis entellus*	55
Colobus polykomos	150	*Presbytis obscura*	62
Colobus abyssinicus	3	*Presbytis francoisi*	25
Cynopithecus niger	115	*Presbytis cristatus*	47
Erythrocebus patas	14262	*Presbytis melalophos*	10
Macaca silenus	65	*Presbytis potenziani*	18
Macaca irus	53432	*Pygathrix nemaeus*	5
Macaca mulatta	39263	*Theropithecus gelada*	22
Macaca nemestrina	1163		
Macaca speciosa	1415		
Macaca maurus	100	Total	147366

TABLE VIII

Imports of Callitrichidae (1965-1975 Inclusive into the United Kingdom

Callithrix j. jacchus	4966
Callithrix jacchus aurita	828
Callithrix jacchus geoffroyi	83
Callithrix pencillata	319
Callithrix argentata	107
Cebuella pygmaea	250
Leontopithecus rosalia	18
Saguinus fuscicollis illigeri	46
Saguinus fuscicollis melanoleucus	50
Saguinus midas midas	16
Saguinus midas tamarin	200
Saguinus mystax mystax	30
Saguinus nigricollis	473
Saguinus oedipus oedipus	979
Saguinus oedipus leucopus	2
Saguinus oedipus geoffroyi	63
Saguinus labiatus	130
Total	8560

TABLE IX

Primates passing through the RSPCA Hostel, Heathrow (London) Airport, 1974

Description	Number	Number Dead
Rhesus	17884	58
Black apes	5	-
Marmosets	57	8
Squirrel monkeys	425	5
Mandrill	5	-
Deprazza monkeys	42	1
Mangabeys	2	-
Putty nosed monkeys	13	-
Diana monkeys	2	-
Howler monkeys	15	1
Guenons	1	-
Baboons	1153	-
Woolly monkeys	4	-
Patas monkeys	248	1
Orang utans	1	-
Capuchins	44	-
Javan monkeys	14	-
Night monkeys	15	1
Mona monkeys	3	-
Gibbons	11	1
Siamangs	2	-
Langurs	16	-
Saki monkeys	11	-
Chimpanzees	15	-
Gorillas	2	-
Unidentified	14	-

not clear that the *Saguinus oedipus* which presumably originated in Colombia, were legally exported, nor is it clear that the other animals were accompanied by export documentation.

Cebidae

A total of 24,035 of 24 species were imported (Table V). From 1965-67 an average of just over 5,000 per annum were licensed. In 1968, the first year when the actual numbers imported were published, 1,543 came into the United Kingdom, and from 1968-75 an average of 1,067 per annum were imported; in 1975, 500 individuals of 7 species were imported, the lowest number recorded.

The most numerous imports were 1,182 *Aotus trivirgatus* (5%), 2,051 *Cebus capucinus* (8.5%), 1,609 *Lagothrix lagotricha* (6.7%) and 15,145 *Saimiri sciureus* (63%).

Cercopithecidae

A total of 147,366 of 45 species were imported (Table VII). As with the cebids the official statistics give unrealistically high figures for 1965-67.

TABLE X

Imports of Primates listed in the IUCN Red Data Book, Mammalia, 1972 Edition

Vulnerable	1965	1966	1967	1968	1969	1970	1971	1972	1973	1974	1975	Totals
Cheirogaleus	-	-	-	-	10	-	-	-	-	-	-	10
Lemur mongoz	1	-	-	-	-	-	-	-	-	-	-	1
Gorilla gorilla	2	2	6	1	4	-	-	16	4	-	4	39
Pan troglodytes	44	23	31	16	46	3	4	12	16	6	1	202
P. paniscus	-	-	1	-	-	-	-	-	-	-	-	1
Endangered												
Pongo pygmaeus	1	3	15	4	1	1	-	-	-	-	-	25
Brachyteles arachnoides	24	-	-	-	-	-	-	-	-	-	-	24
Cacajao rubicundus	-	-	-	5	1	-	-	-	1	-	-	7
Leontopithecus rosalia	6	6	6	-	-	-	-	-	-	-	-	18
Macaca silenus	50	2	2	-	11	-	-	2	-	-	-	65
Pygathrix nemaeus	-	-	-	-	-	-	-	-	-	3	-	5

From 1968-75 the average was 10,551 per annum. The most frequent imports were 11,323 *Cercopithecus aethiops* (7.7%), 14,262 *Erythrocebus patas* (9.7%), 53,432 *Macaca fascicularis* (36.3%), 39,263 *Macaca mulatta* (26.6%) and 11,940 *Papio papio* (8.1%).

From the table it will be seen that in spite of the fact that imports of certain species have declined drastically, the annual totals have been maintained by increased imports of other species. For instance, *Cercopithecus aethiops* fell from 974 in 1968 to 25 in 1975, *Erythrocebus patas* from 1,275 to 117 but *Macaca fascicularis* rose from 3,574 to 4,725. It is noticeable that it is the African cercopithecid and the cebid imports which have fallen, while the imports of Asian cercopithecids have stayed more or less stable, or even risen.

TRANSIT ANIMALS

In addition to the animals actually imported into the United Kingdom, there are large numbers passing through. Table IX gives a summary of the primates recorded at the RSPCA Animal Hostel. Unfortunately the species are not always identifiable, but enough are to make some sort of assessment. In research on behalf of the Royal Society for Protection of Birds, T. Inskipp (personal communications) found that about 25% of birds passing through the RSPCA Hostel were actually imported into the United Kingdom, the rest continuing to the United States and other destinations. He also estimated that only 10-20% of birds imported into the United Kingdom pass through the RSPCA Hostel. Clearly this does not hold true for primates, as over 60% were destined for the United Kingdom. Studies at the RSPCA Hostel have also led to the discovery of several consignments of animals being smuggled into the United Kingdom. More recently it has been found that some animals may be entering the United Kingdom without the documentation which is required for an import licence to be issued in compliance with the Washington Convention in the Trade in Endangered Species (1973).

It is important, however, to put these cases in a broad perspective; while it is important to prevent evasion of the law, it is evident that the import into the United Kingdom of the higher primates (Table VI) and rarer species (Table X) is on a very small scale, whereas the import of Asian cercopithecids continues unabated - in fact under the recent changes in British law certain species did not require licences for a period between the changes in legislation.

SUMMARY

The United Kingdom restrictions governing the imports of live primates are generally effective, and ensure that adequate protection is given to endangered species. However, the broad trends shown by an examination of the imports from 1965-75 inclusive indicate that as one species is depleted (and consequently protected) the demand shifts to others. It is essential therefore that all primate imports are strictly monitored and that import quotas are set. In order to prevent the demand shifting, quotas should never be set higher than the average recorded import over the past decade unless good estimates of the harvestable population and the effects of such a harvest are available.

ECONOMICS AND DISEASE IN IMPORTED SIMIAN PRIMATES

G.W. TRIBE AND D.A. BASSETT

*Shamrock Farms (GB) Ltd, Upper Horton Farm,
Small Dole, Henfield, Sussex UK.*

INTRODUCTION

The concept of a relationship between economics and disease is not new. The Office of Health Economics has published a series of articles showing the relationship between economic loss and such diverse conditions as medical care (Anon., 1964), chronic bronchitis (Anon., 1963a) and venereal disease (Anon., 1963b), while there is a large bibliography relating to such topics as the economic consequences of calf scours (Loosmore, 1970), neonatal sheep mortality (Pout et al., 1973), bovine helminthiasias (Michel, 1976) and swine fever (Ellis, 1972). Clearly, there is a relationship between the incidence of disease amongst captive simian primates and the cost of these animals to the research worker, although it has not hitherto been defined in detail.

BASIC COST ELEMENTS

The cost of a conditioned simian primate is determined by seven different factors. Four of these are not significantly affected by disease amongst the animals. For example, the initial purchase price and cost of freight to Great Britain is set by agents in the country of origin and commercial airlines, while labour costs are determined more by the demands of good husbandry than by problems arising out of incidence of disease. Similarly, general overhead charges and administrative, selling and distribution costs are not related, but revenue devoted to purchase of feed stuffs and medication clearly is. Veterinary costs are incurred not only by the demands of treatment regimes in the face of disease but also by the necessity for carrying out extensive health screening before releasing animals, to which must be added the cost of laboratory services.

VARIABLE COST ELEMENTS

Variable cost elements are related to deaths, batch turnover and occupancy rate. Batch turnover may be defined as the length of time taken to dispose of an entire batch from the date of importation, while occupancy is the proportion of cages containing an animal compared to empty cages within a unit. Maximum efficiency demands a turnover period of 11 weeks with 100% occupancy. Bassett (1976) has shown there is an inverse relationship between profitability and increased periods of conditioning and it is obvious that labour and general overheads will remain roughly the same whether a unit is full or half empty. Thus where disease either extends conditioning periods or reduces the

THE DISEASE SPECTRUM AND OPERATING EFFICIENCY

Transit deaths

First evidence of disease amongst a group of imported animals is often the bodies of those which died en-route. Fortunately, such losses are extremely low but they do have a bearing on batch economics because animals are imported from overseas in numbers corresponding exactly to vacant cages. Hence, where transit losses occur, occupancy rates fall immediately and economic efficiency is impaired.

Surgical conditions

Imported monkeys are prone to a variety of surgical conditions including fractured limbs, abscesses, lacerations, dental disease, gangrenous tails, joint infections and gross loss of skin and muscle tissues. Most of these respond to treatment in a remarkably favourable manner and these conditions do not have any significant bearing on the economics of the enterprise. However, one rather sad syndrome is the occasional arrival at our premises of monkeys with one leg, one eye or distorted spines. Biomedical research has no use for such animals. It is a tragedy they have made the long journey from Africa or Asia only to suffer humane destruction on arrival at the quarantine premises. In terms of efficiency, however, there are two effects. Firstly the occupancy rate is reduced and secondly, allocation of animals to orders is somewhat disorganised.

Effects of malnutrition

When animals arrive from overseas they are frequently in a poor nutritional condition. Worst affected are baboons, but often cynomolgus monkeys show a range of deficiency diseases resulting from a combination of heavy worm burden and dietary insufficiency.

Evidence of deficiency disease is readily obtained by simple physical examination. Muscle wasting, lack of periorbital fat, poor coat and swollen epiphyses tell their own tale. Vitamin and mineral imbalances are readily corrected by use of modern pharmaceutical preparations but the protein deficiency is rather a different matter. Not only must a diet containing adequate levels of protein be fed, but the protein contained therein must be of a high quality. Basically, this means that not only should it contain only vegetable matter, thereby avoiding rancidity and deterioration of essential fat-soluble vitamins, but there should be adequate amounts of essential amino-acids including thiamine, arginine, lysine and leucine. Such diets are readily obtained but are much more expensive than those suitable for maintenance of stabilised animals.

We have found that provided such a diet is fed consistently over the first four weeks of holding and that the animals are freed from helminths shortly after arrival, the effects of dietary insufficiency soon disappear. Indeed the transformation in baboons is truly remarkable and we find we can change these animals to a maintenance diet after intensive care for one month.

Despite the high cost of top quality feed, considerable advantages result from its use. For example, in trials done by comparing the effects of superior and inferior quality feeds within groups of cynomolus monkeys, deaths were reduced by 30%, response to treatment in disease was remarkably improved and the number of animals failing the clinical examination after eight weeks conditioning drastically curtailed in those groups receiving the better quality feed (Tribe and Welburn, unpublished information). Hence, despite very high outlay on feed, the efficiency and hence profitability of the enterprise may be improved through increased turnover and occupancy.

Helminthiasis

Infestations with a variety of parasitic worms are commonplace. Ruch (1959) gave a comprehensive description of the types which can occur, while Owen and Cassillo (1973) published the results of a survey of animals newly-arrived in Great Britain prior to commencement of treatment. From the economic point of view most serious are *Oesophagostomum* species and echinococcal cysts. The former are commonplace in all species imported by us and, although readily dealt with by parenteral levamisole, the larval forms frequently induce peritonitis. This means that the animal remains an invalid for some weeks, particularly where macaques are involved. Frequently this means such animals outstay their welcome and although they may eventually recover fully they represent an economic loss solely because of the length of time they must be kept.

Echinococcal cysts are seen infrequently but where they do occur they represent a major surgical problem. When they are lodged in the chest cavity the condition is usually diagnosed as tuberculosis from the radiograph and the animal killed, the true nature of the lesion only becoming evident at autopsy. Subcutaneous cysts can be removed surgically but in our experience several operations are invariably necessary to rid the animal of the lesion completely. Consequently, batches containing such animals have an extended turnover period.

Pneumo-enteritis

Of all the treatments administered at our unit probably 80% are given because of enteritis. Some 15% of the remainder are accounted for by pneumonia outbreaks. The latter are most dangerous when they occur in newly-arrived animals because often the lungs are grossly consolidated and these patients are really beyond treatment. Fortunately, such incidents are now rare although respiratory infections are commonplace during conditioning. Such illness invariably responds uneventfully to modern therapeutic agents and apart from the cost of the drugs, which is insignificant, does not really have much economic impact.

Enteritis, often associated with helminth-induced trauma allied to malnutrition, can be a serious economic problem, however. It is responsible for approximately 80% of the deaths which occur during conditioning and is related to infection with a variety of bacteria, most notable of which are *Salmonella* and *Shigella* species. Cases frequently require long and tedious treatment regimes which are expensive in terms of manpower. More serious from the economic point of view is the fact that seemingly healthy animals will on occasion develop dysentery two or three days prior to the time of despatch, which usually means the departure of the entire order has to be delayed. Thus, the client's time-table is disrupted often with gross inconvenience and financial loss to him, certainly with financial loss to us, because the turnover factor is adversely affected.

The economic effects of simian tuberculosis

Simian tuberculosis tends to occur sporadically in newly-imported monkeys, particularly in baboons and rhesus. The cost of controlling this infection falls under two heads. First, there are losses associated with deaths or with animals which have to be killed on the basis of positive screening test results. Secondly, there is the cost of carrying out the surveillance programme, and of these two it is possible the cost of the latter exceeds that of the former. This is because although only 2% of animals are infected, the screening programme must be conducted on the assumption that any imported primate is a potential carrier of tuberculosis.

Nevertheless, where a case of tuberculosis does occur, the economic effects are out of all proportion to the value of the individual animal involved. This is because we have to go to great lengths to ensure other animals in the batch have not contracted infection, which obviously extends the conditioning period thereby reducing turnover. Occupancy is little affected nowadays because improved methods of tuberculosis detection have virtually abolished the false positive syndrome which hitherto caused many animals to be killed unnecessarily (Tribe and Welburn, 1976; Acred et al., 1972).

It is salutary, however, that a batch of 50 baboons will require a total of 15 man-hours to carry out an effective screen against tuberculosis in a batch which is free of the disease. Where tuberculosis does occur this figure may have to be trebled, and much of this time has to be paid for at professional rates.

The impact of moderately pathogenic viruses

An example of a moderately pathogenic virus is simian varicella (Almeida, McCarthy and Tribe, unpublished information). This agent, a member of the herpes group of viruses, rapidly infects all animals in a batch, causes extensive vesiculation in about 10% with petechial haemorrhages throughout the viscera. The vesicular fluid contains a mass of virions which undoubtedly cause widespread contamination when they rupture. Besides the rash, which is obvious in only a minority of animals, the agent also induces pneumo-enteritis.

Animals displaying an extensive rash invariably die. Losses vary from outbreak to outbreak but usually between 20% and 50% of the batch will be lost. Such losses tend to occur in the first four weeks of holding, which means that units have to remain half empty for at least four weeks with consequential effects on the occupancy factor. During such an epidemic, animals must receive intensive care and it is necessary to give antibacterial therapy on a daily basis to avoid additional losses from bacterial pneumonia. Hence the financial consequences of such outbreaks are serious and because, on average, about one batch in 15 is infected with an agent of this order of pathogenicity, such incidents cannot be ignored when costings are prepared.

The shock of a peracute lethal infection

Probably the most spectacular agent in this category is that responsible for the Marburg disease syndrome (Smith et al., 1967) because it not only kills animals it infects but frequently kills staff in attendance. We have not knowingly seen this virus at our unit but we have experienced five outbreaks of an infection known as Sukhumi disease, which is lethal to macaques but does not, fortunately, affect man. Hitherto, this agent totally destroyed monkey colonies in the USA, Russia and Japan. In our first three epidemics only one animal survived from all the batches affected, but in the penultimate outbreak 75% of the batch was saved by dividing the batch into small subgroups and housing these separately. On that occasion, where subgroups became infected all the animals therein died. In the last outbreak (1976) the disease was suspected when the first animal was taken ill and the most valuable animals were moved into small isolation units. All these were saved. The residue were given no treatment whatsoever but animals becoming affected were killed immediately. Many of these displayed epistaxis and the classical large petechial haemorrhages in viscera reminiscent of classical swine fever lesions. This technique was instrumental in saving approximately 50% of the animals remaining in the affected unit. However, it was fortuitous that an early diagnosis was made so that control measures could be put into effect immediately.

Although Sukhumi disease of macaques is the best known agent of this type, it should be remembered that other haemorrhagic fevers do occur. For example one of us (G.W.T.) has seen entire batches of *Erythrocebus patas* and marmoset monkeys destroyed by agents of this type, and such incidents must of necessity increase the overall cost of a conditioning enterprise.

CONCLUSIONS

Certain elements in the cost of a conditioned simian primate, such as the price paid to the trapper in the country of origin and the air freight, are beyond the control of the conditioning agency. Moreover, items such as labour costs are bound to be high and are unavoidable. Hence, the only variable elements are those of mortality, occupancy and batch turnover. Disease only affects profit levels when it leads to deaths and reduction of occupancy or extends conditioning periods. The length of the prescribed conditioning period will depend to a large extent on the policy of the people responsible, who will use their experience and judgment to determine when animals are likely to be ready for despatch to the laboratory. However, those using such animals in Great Britain are bound to take note of the implications of the Health and Safety at Work etc. Act (1974). Although an eight week conditioning period only amounts to one third of the statutory six months quarantine demanded by the Rabies (Importation of Dogs, Cats and Other Mammals) Order (1974) it has long been accepted in Great Britain that simian primates should be conditioned for a minimum of eight weeks before going to a laboratory where experimental work is carried out. Hence, if they are taken earlier and an incident occurs leading to litigation, the court might well hold that the governing authority of the laboratory was negligent in law. Hence, although a two week conditioning period would give an apparently cheaper animal and would benefit the conditioning agency because many more stockturns would be possible, such action would in the long term benefit no-one.

Clearly, control of disease is of paramount economic importance as well as being desirable for humanitarian reasons. A disease control programme must aim at reducing mortality and preventing spread of infections because the former reduces occupancy and the latter increases length of conditioning periods. Economic success or failure will depend ultimately on how effectively this task is discharged.

SUMMARY

The relationship between the cost of a conditioned simian primate and disease is a reflection of mortality, unit occupancy and efficiency of stock turnover. Whereas some diseases such as pneumo-enteritis, which can be controlled by therapy, do not have much influence on costings, agents which kill a large proportion of the animals obviously do. Diseases such as tuberculosis, however, although they can be controlled by modern techniques, increase costs because the length of conditioning periods are extended. Thus overall turnover is reduced while overheads remain the same. Therefore disease control must not only be directed at reduction of mortality but also towards keeping occupancy high and turnover at predicted levels.

REFERENCES

Acred, P., Hunter, P.A., Bywater, J.E.C. and James, J. (1972). *In* "Medical Primatology" (W.I.B. Beveridge, ed.). S. Karger, Basel.
Anon. (1963a). "Pneumonia in Decline" (*ibid.*)
Anon. (1963b). "The Venereal Diseases" (*ibid.*)
Anon. (1964). "The Costs of Medical Care". Office of Health Economics Publication, London.
Bassett, D.A. (1976). *Primate Supply*, $\underline{1}$, No. 2, 8-9.
Ellis, P.R. (1972). An economic evaluation of the swine fever eradication programme in Great Britain. University of Reading, Department of Agriculture.
Health and Safety at Work Act (1974). HMSO, London.
Importation of Dogs, Cats and other Mammals Order (Prevention of Rabies) (1974). HMSO, London.
Loosmore, R.M. (1970). "Calf Mortality - the problem". Agricultural Development Association, York.
Michel, J.F. (1976). *ADAS Q. Rev.*, $\underline{20}$, 162-177.
Owen, D. and Casillo, S. (1973). *Lab. Anim.*, $\underline{7}$, 265-269.
Pout, D.D. and Thomas, W.J.K. (1973). Veterinary and medicine costs and practices in lowland sheep. University of Exeter, Agricultural Economics Unit.
Ruch, T.C. (1959). *In* "Diseases of Laboratory Primates"

Smith, C.E.G., Simpson, D.I.H., Bowen, E.T.W. and Zlotnik, I. (1967). *Lancet*, $\underline{2}$, 1119.
Tauraso, N.M., Myers, M.G., McCarthy, K. and Tribe, G.W. (1970). *In* "Infections and Immunosuppression in Subhuman Primates". Munksgaard, Copenhagen.
Tribe, G.W. and Welburn, A.E. (1976). *Lab. Anim.*, $\underline{10}$, 39-46.

COST ANALYSIS AND RATE SETTING IN A PRIMATE ANIMAL LABORATORY

T.A. FITZGERALD

*Office of Grants Administration and Institutional Studies,
New York University Medical Center,
New York, USA.*

INTRODUCTION

Within the past decade, increased use of non-human primates in biomedical research, coupled with reduced financial support of research, led to a need for improved fiscal accountability and rate setting. For this purpose a model was first developed for the Laboratory for Experimental Medicine and Surgery in Primates (LEMSIP) of the New York University Medical Center and was presented at the Third Conference on Medical Primatology in Lyon (Davis and Fitzgerald, 1972). The economic theories and methodology defined were later applied to the development of a model for animal resource facilities generally, which was intended to:
1. Help animal facilities become self-supporting by recovery of costs from users.
2. Educate users in the true cost of animal experimentation.
3. Enhance the quality of animal care and the research and educational functions that depend upon animals.
4. Document animal costs accurately in grant and contract proposals.
5. Ensure that grant and contract awards are sufficient to cover animal costs.
6. Improve the efficiency of animal management.

Development of Cost Analysis and Rate Setting Manual

Accordingly, the United States Department of Health, Education and Welfare, National Institutes of Health, Division of Research Resources - Animal Resources Branch, with the Association of American Medical Colleges and its Group on Business Affairs (GBA) established a joint committee called the Division of Research Resources Ad-Hoc Committee on Animal Costs. The work of the committee produced the *Cost Analysis and Rate Setting Manual for Animal Resource Facilities* (DHEW, NIH, 1974). In developing the manual, the committee drew heavily upon the above mentioned *Fiscal Management of a Primate Facility for Biomedical Research* (Davis and Fitzgerald, 1972), and *Cost Finding and Rate Setting for Hospitals* of the American Hospital Association (1968). The model was field tested in nine universities selected on the basis of the following criteria:
1. The animal unit should be a central facility serving at least an entire medical school. The director should have an interest in and capability of conducting the test to a successful conclusion.

TABLE I
Comparison of Functions

The Primate Facility	The Hospital

Cage complement
  By type of animal
	Bed complement
	  By type of service

Building services
  Operation and maintenance of plant
  Housekeeping services
	Building services
	  Operation and maintenance of plant
	  Housekeeping services

General and administrative services
  Professional care administration
  Business services
  Medical records and statistics
  Communications
  Employee health care
	General and administrative services
	  Professional care administration
	  Business services
	  Medical records and statistics
	  Communications
	  Employee health care

Dietary service	Dietary service
Laundry service	Laundry service
Transportation service	Ambulance service

Clinical laboratories
  Hematology
  Pathology and histology
  Microbiology
  Chemistry
	Clinical laboratories
	  Hematology
	  Pathology and histology
	  Bacteriology
	  Chemistry, etc.

Surgery	Surgery
X-Ray	X-Ray
Pharmacy	Pharmacy
Teaching programs	Teaching programs
Research programs	Research programs

Animal husbandry services
  Professional services
    House staff - DVM
 	
  Nursing services
    Senior technicians
    Technicians
    Junior Technicians
	Patient care services
	  Professional services
	    House officers
	    Interns and residents
	  Nursing services
	    Registered nurses
	    Licensed Practical Nurses
	    Nurses' aids

  2. The university should have a business office with the interest and capability of participating in the test.
  3. There should be diversity among the participating animal facilities in order to test the flexibility and suitability of the model in the widest

TABLE II

Step 1: Trial Balance
List all Expenses Direct and Indirect

Direct Expenses

Salaries
Wages
Related fringe benefits
Moveable equipment
Supplies and materials
Dietary and bedding
Service contracts
Travel
Other

Indirect Expenses

Allocated share of:
  Plant operations
  Library operations
  Administrative and general expense

Use charge for:
  Equipment
  Buildings etc.

Total Direct and Indirect Expenses

Animal procurement:
  Cost of animals
  Plus: Facility service fee

Total

---

variety of situations. For example the test included large and small, public and private, new and old animal facilities with a wide geographical distribution.

The pilot test disclosed problems inherent in both the model and the individual facilities. Because laboratory services, animal health care and research services are areas in which operations tend to overlap, it was necessary in the revision of the model to define these accurately.

(1) Laboratory services

All costs associated with the maintenance and operation of support laboratories such as those dealing with hematology, pathology, clinical biochemistry and bacteriology.

(2) Animal health care

The diagnosis, treatment and prevention of disease and injury and the maintenance of the good health of the animals.

(3) Research services

Technical services provided in the execution of a research protocol as

TABLE III

Step 2: Assignment of Expenses Listed in the Trial Balance to Facility Cost Centers

Cost Centers

1. Operation and maintenance of plant
2. Facility administration
3. Transportation
4. Cage washing and refuse disposal
5. Dietary and bedding
6. Laboratory service
7. Animal health care
8. Surgical and X-ray services
9. Research services
10. Animal husbandry:
    a. Small new world monkeys
    b. Medium size new world monkeys
    c. Small old world monkeys
    d. medium size old world monkeys
    e. Adult chimpanzees
    f. Juvenile chimpanzees
    g. Nursery
    h. Quarantine

ordered by an investigator.

Another problem involved identifying and establishing a reasonable service fee for animal procurement in institutions where the animal facility has full responsibility for this. Although the model provides a method for determining a fee, it may need modification. As more institutions use the model and gain better insight into their costs, they will be able to suggest improvements. Absence of detailed accounting data was another difficulty encountered by institutions testing the model. Existing charts of accounts did not record income and expenses in a fashion which would allow completion of the study in accordance with recommended methodology. It was necessary, therefore, to analyse existing data in order that expenses could be assigned to appropriate cost centers. As a result of these first field tests the participating universities are improving their accounting systems, which will facilitate future studies. Further, several of these universities are establishing computer based programs of accounting and statistical data retention that will make the model operational on a permanent basis.

This will help overcome the most significant problem that arose during the pilot test, namely the lack of accurate statistical data. Hospital managements have long appreciated the value of such data in the calculation of rates for services provided. Hospitals have, therefore, developed refined systems for capturing and retaining statistical data.

Another valuable by-product of this pilot study was the recognition by participating universities and institutions that data-keeping would have to be improved.

Methodology provided is easily adaptable to the different types of research animal environments without compromising professional standards. Depending on the needs of the institutions, the cost centers were suitably modified.

TABLE IV

Step 3: *Allocation of Non-Revenue Producing Cost Centers to Revenue Producing Cost Centers*

Non-Revenue Producing Cost Centers

1. Operation and maintenance of plant
2. Facility administration
3. Transportation
4. Cage washing and disposal
5. Dietary and bedding
6. Laboratory services
7. Animal health care

Revenue Producing Cost Centers

8. Surgical and X-ray services
9. Research services
10. Animal husbandry:
    a. Small new world monkeys
    b. Medium size new world monkeys
    c. Small old world monkeys
    d. Medium size old world monkeys
    e. Adult chimpanzees
    f. Juvenile chimpanzees
    g. Nursery
    h. Quarantine

The Cost Analysis and Rate Setting Manual for Animal Resource Facilities outlines a method for identifying all direct and indirect costs, the assignment of expenses to their proper cost centers and the allocation of costs from non-revenue to revenue-producing cost centers.

Effective management should be as concerned with the goodwill of users as it is with costs and their recovery. Use of accurate data as outlined in the model should enable (i) identification of spending patterns by cost centers for the purposes of constructing a budget; (ii) improved fiscal planning; (iii) a permanent budgeting mechanism for current and future operations; (iv) justification of costs and related fees to principal investigators, sponsors and administrators.

The methodology prescribed in the manual is based primarily on cost accounting procedures and methods used by American hospitals. There is a similarity between the functional operation of hospitals, primate resource and general animal resource facilities. A comparison of functions is illustrated in Table I. Respecting cost analysis and rate setting it recommends a study in three major steps:

1. The identification of all direct and indirect expenses incurred in the facility's fiscal year.
2. The assignment of these expenses to functional cost centers.
3. The charging of expenses to revenue producing rather than non-revenue producing cost centers.

These data provide a base for the development of a unit cost for each revenue producing service which is necessary for rate setting.

Three prerequisites for cost analysis are sound organizational structure, an appropriate accounting system, and adequate statistical data. A sound organizational structure provides lines of authority and responsibility. It should allow for sub-division of the primate resource facility into functional units. Conduct of a study should, therefore, be undertaken as a team effort. The majority of animal resource facilities are incorporated in the parent institution's total accounting system. These systems are usually designed to account for many activities, including education, research and patient care. Any cost accounting system, designed for animal resource facilities must, therefore, receive data from the parent institution's fiscal and accounting personnel. Application of adequate statistical data is essential for the assignment of expenses to cost centers and the re-allocation of expenses from non-revenue producing cost centers. Adequate statistical data must also be available for development of unit costs for rate setting purposes.

When cost-analysis procedures are introduced into an animal resource facility, it is unlikely that all essential statistical information will be available. Reasonable assumptions and substitutes must therefore be devised for the first study, but programs ought to be initiated for the collection of data for use in the future.

CONDUCT OF A STUDY

1. Identification and recording of all direct and indirect expenses incurred during the fiscal year, for example, in the accountant's terminology, "preparation of the trial balance". Table II illustrates major items of expense included in the trial balance, e.g. salaries, wages and supplies.

Recording of actual expenses must be done in sufficient detail to allow for assignment to relevant cost centers.

2. Assignment of expenses appearing in the trial balance to functional cost centers (Table III). There are ten of these but animal husbandry is sub-divided for eight species of animal. Cost centers as illustrated in Table III may be modified in accordance with the specific needs of individual facilities without compromising study methodology. If the cost analysis program is to be successful, management must carefully consider identification of cost centers before starting the study. Once pertinent cost centers are identified, expenses listed in the trial balance may be assigned to them.

Figure 1 is an illustration of a typical work paper, in summary form. Assignment of expenses may be accomplished by either direct identification

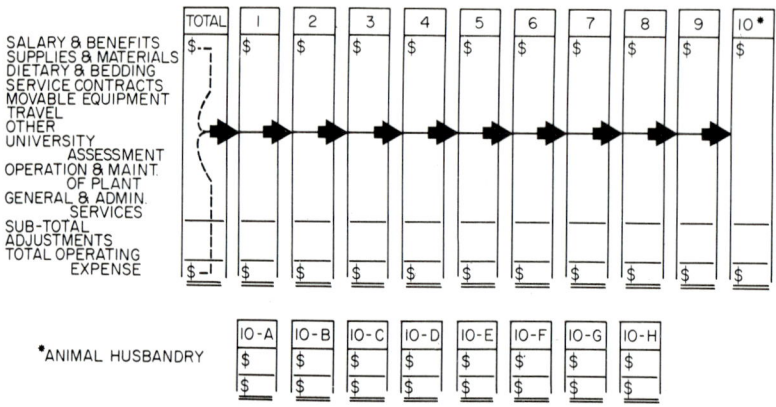

Figure 1. *Facility Cost Centers*

with a specific cost center or by apportionment to a number of centers using the relevant statistical information for that particular type of expense. Salary and wage expense incurred for animal husbandry personnel may, for example, be assigned directly to the cost center animal husbandry. To break down this expense as a function of species it is necessary to apply statistical data which may be obtained from individual time and effort reports. Further, time and motion studies should be conducted periodically, which may be limited to periods of one working week three to four times annually.

3. The allocation of costs from non-revenue to revenue producing cost centers (Table IV). Cost centers identified as operation and maintenance of plant and facility administration are not revenue producing functions and must, therefore, be allocated to revenue producing functions. Laboratory services, however, may be considered in either category. If the facility operates the laboratory purely in support of the animal husbandry function, the cost center would be considered non-revenue producing, and allocated to the animal husbandry function. Should the laboratory be operated for the provision of services to principal investigators, the cost centers should be considered revenue producing.

The re-allocation of costs in the stage 3 is accomplished via the step-down method. This allows for the allocation of certain non-revenue producing cost centers to others than to revenue producing cost centers. To facilitate this it is necessary to prepare appropriate work paper. The matrix is illustrated in Figure 2. It will be noted that non-revenue producing cost centers appear in the horizontal input of the matrix, while both the non-revenue producing centers appear in the vertical.

The order of presentation of cost centers in the matrix gives consideration to the fact that certain non-revenue producing functions service other non-revenue producing functions; therefore, the allocation of these functional cost centers is handled in the stepping down procedure.

Here again, the application of adequate statistical data is essential if accurate results are to be achieved.

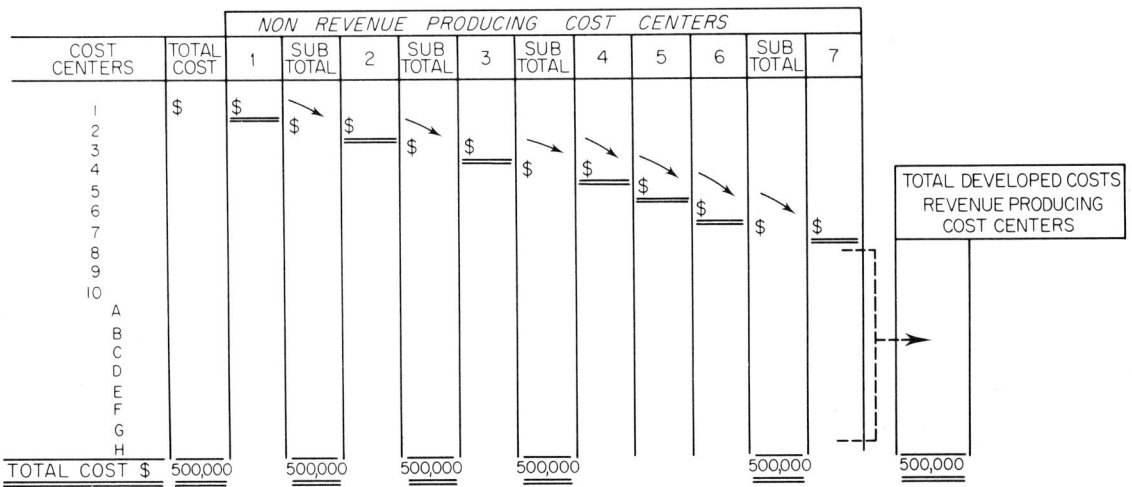

Figure 2. *The step down.*

The first cost center to be allocated under the step-down method is that for the operation and maintenance of plant. This may be done by calculating the cost per square foot and applying this figure to the space assigned to the remaining cost centers.

The second is facility administration. This may be accomplished by applying total administrative cost as a function of total accumulated cost for each of the remaining cost centers. The same approach may be applied to the handling of cost center 3 (transportation) and cost center 7 (animal health care). Cost center 4 (cage washing and disposal) may be allocated, based on number of cages multiplied by average frequency of washing per week.

Annual feed and bedding costs may be allocated on the basis of price multiplied by quantity per animal in typical units during a sample period multiplied by the number of animal days per year. Laboratory services may be allocated on the basis of cost divided by the number of tests performed. Tests performed should be valued on the estimated average time per test.

Completion of this final step for cost analysis provides the study with the total developed cost for each revenue-producing function. These data, coupled with the application of appropriate statistics, will allow for calculation of unit cost for each revenue-producing service.

Samples of the application of statistical data for the development of unit cost are:

1. Surgical and X-ray services - developed cost as a function of service units equals cost per service unit. A service unit may be considered as an hour or any appropriate unit of time.

2. Research services (technical) - same approach as described for surgical and X-ray services.

3. Research services (laboratory test) - developed cost taken as a function of the total number of tests performed equals cost per test. Tests may be weighted to allow for time differentials inherent in the different types of tests.

4. Animal husbandry - developed cost by species taken as a function of total annual animal days equals cost per day.

5. Animal procurement - developed cost taken as a function of animals assigned equals cost per animal. Animals assigned may be considered as the number of animals purchased less the number of animals lost during the initial quarantine period.

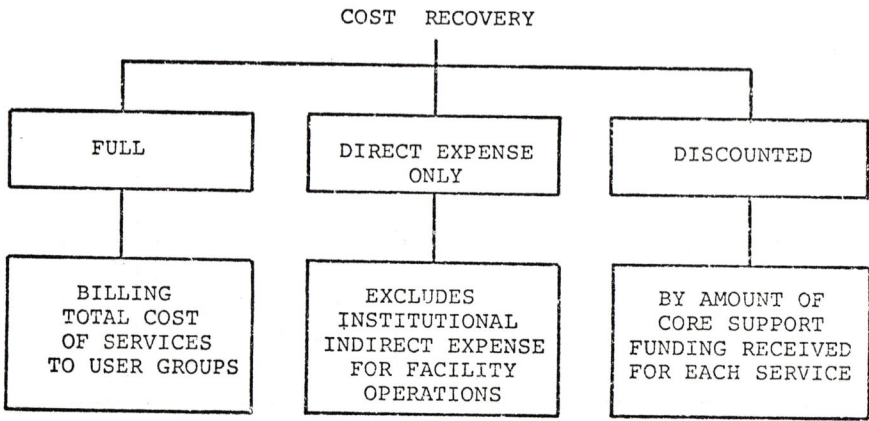

Figure 3. Rate setting.

Rate setting involves the method by which the facility expects to recover its cost of operations. Three approaches of cost recovery are illustrated in Figure 3. The first assumes the recovery of full cost. Accordingly, rate setting is equal to the developed unit cost for each service. The next method assumes recovery of only direct cost, excluding any consideration of the parent institution's cost for the operation and maintenance of plant, and other institutional administrative services. This type of handling is only applicable in cases where the parent institution has the capability to fund such costs through other resources. The final consideration of cost recovery involves discounting rates established under either of the two methods discussed above, by the application of core support funding made available by other sources. In this respect, should the facility have the benefits of an endowment fund supporting, for example, its surgical and X-ray suite, income earned from the endowment would be applied to the developed unit cost of service in order to establish the discounted rate. To the non-accountant, the first reading of the manual may suggest that implementation of the economic model in practice is difficult, but if undertaken in a logical, orderly way, it is not. Experience shows that a major pitfall is the tendency to over-identify and define in the initial phase. This can be easily avoided through a thorough understanding of the manual and the clear establishment of study goals.

The manual has been popular beyond expectation and is now in second printing. Copies have been requested from institutions throughout the world. Many of them have now completed pilot studies and have established programs to conduct the study on a regular basis.

A plan is presently under consideration to revise methods and procedures described in the manual. The plan calls for revising and adding to methodology applicable to indirect expense, statistical record-keeping, and computerization of the economic model. If this plan is adopted, it is anticipated that a revised printing will be available in the latter part of 1977.

REFERENCES

American Hospital Association (1968). "Cost Finding and Rate Setting for Hospitals". American Hospital Association, Chicago, Illinois.

Davis, Jr., and Fitzgerald, T. (1972). *In* "Medical Primatology" (E.I. Goldsmith and J. Moor-Jankowski, eds), Part 1, pp.37-49. Karger, Basel.

DHEW, NIH and Association of American Medical Colleges - Division of Research Resources (1974). "Cost Analysis and Rate Setting Manual for Animal Resource Facilities". NIH, Washington.

SECTION III

BREEDING PRIMATES IN CAPTIVITY
(W.R. Dukelow and W. Lane-Petter, eds.)

TUPAIA BELANGERI BIAS, AN OUTBRED STOCK OF TREE SHREWS

ANITA SCHWAIER

Battelle-Institut eV, Frankfurt/Main, Germany.

Numerous statements have been published by international organisations such as WHO on the urgent necessity not only of breeding primates for medical research, but also of investigating possible ways of replacing precious primate species in experimental research by species whose existence is not endangered and which are available in large numbers. Our tupaia colony was started with just this objective in mind.

TABLE I

Breeding Progress of the Tupaia Colony. Each year one new generation is added to the stock replacing the oldest, which is eliminated

Generation	Animals Surviving the Weaning Age				
	1972	1973	1974	1975	Jan-Jun 1976
F1	102	54	2	6	-
F2	9	60	68	32	4
F3	-	3	52	109	47
F4	-	-	14	56	53
F5	-	-	-	15	30
F6	-	-	-	-	18
F7					
Total	111	117	136	218	152
Mean Litter Size	2.1	2.4	2.5	2.4	2.7
Survival Rate %	53	61	62	56	52

We can look back on five years of experience in breeding tupaias. In the course of the present year we shall wean Animal Number 1000 of our colony (Table I). As we apply a defined mating system, criteria for an acknowledged stock according to the rules of the International Committee for Laboratory Animals are fulfilled. The full name of our line is *Tupaia belangeri* Baf: BIAS, where Baf designates the breeder - Battelle Frankfurt - and BIAS is the name of the stock (Figure 1).

What are the essentials of breeding tupaias? The secret of success is neither a matter of food quality - tupaias feed on almost everything edible from spiders to chocolate without trouble - nor is it a matter of stress sensitivity

Figure 1. Tupaias from the Battelle stock may become quite tame when handled frequently.

in the usual sense: tupaias endure large differences in room temperature, humidity or light, they do not die or fall ill in reaction to noise, blood withdrawal, prolonged restraint in metabolism cages, operations or daily handling. The secret of breeding tupaias is understanding their behaviour.

Breeding experience had to be gained in two spheres: management of the breeders on the one hand, and management of the juveniles from weaning to puberty on the other. Only properly raised animals will grow up to be good breeders.

Information on the behaviour and successful breeding of captured adult tupaias have previously been published by Sprankel (1961), Martin (1968) and van Holst (1969). Their experience was taken into account in our work. Under captive conditions tupaias have to be kept single or in pairs in a cage. We equipped the breeding cages according to the behavioural needs of the pairs, offering them hiding places, resting places, a separate nest and sleeping box, objects for scent marking, a branch for defaecation and the like (Schwaier, 1975).

Severe difficulties, however, were encountered in raising the juveniles.

Young tupaias grow up very quickly. Their weight at birth is around 16 g including 3 to 4 g mother's milk. They double their weight within 6 days and leave the nest around the 25th day with a body weight of about 90 g. In the restricted space of a cage like ours the babies have to be removed before the next litter is born, which is usually 44 days after the previous birth. We separate the babies at between 35 and 40 days. Groups of 4 to 7, mostly 6, weanlings called the Kindergarten, are accommodated in breeding cages and remain together until the end of their growth phase and the beginning of sexual maturity - usually two months later.

Our first groups of weanlings presented us with some unpleasant surprises: in about 50% of all Kindergartens we observed severe fighting among the weanlings. As early as in the first week after weaning some animals behaved unbelievably aggressively towards the other group members. They chased and attacked them heavily, often sitting next to the entrance hole of the tube that leads to the sleeping box, waiting there for the next victim. I want to emphasise again: these little beasts had just been weaned, some had mother's milk in their stomach, they had milk dentition, and their testes were tiny (< 0.1 g). Their attacks resulted in weight loss or only minimal weight increases in all animals, usually also in the despot, in squealing, wounds and especially in neurotic behaviour of the attacked animals. These did not dare to enter the cage, urinated in the sleeping box, carried their food down to the box and struggled in the tubes. When outside, they rushed to and fro or turned stereotyped somersaults.

The aggressive behaviour of the despots did not abate in the course of time, but became even worse.

The juvenile aggression occurred in both sexes, in males five times more often than in females.

If the aggressive animal was separated soon, within one or two days, the remaining group grew up very well. All animals stayed peacefully together without possessiveness towards food, and slept and rested in close contact.

What could be the reason for this juvenile aggression? I first assumed that the time of separation from the family might be too early. Young tupaias leave the nest between the 25th and 30th day and from that time they sleep and rest together only with their parents. Our scheme would concede them only five or ten days of contact with their parents. Extreme fearfulness and aggression was observed in animals kept single during juvenile age. To avoid a possible deprivation syndrome, we left the young with the family group until they were two months old. But aggression then seemed to be even worse, at least more frequent. One animal was left entirely with its parents, and at the age of three months it started to attack them.

I then tried to use one very peaceful couple as foster parents for the weanlings. The female of that couple did not get pregnant for unknown reasons. No juvenile despots occurred in such foster families, but the foster father tried to copulate with the little females and chased them continuously. Other adult foster animals attacked the foreign juveniles, so I gave up this method.

Another hypothesis was that the juvenile aggression might be elicited by the strange smell of the cage mates. Adult female tupaias for instance, are highly selective with respect to their sexual partners. They may attack one male heavily and accept another one. To prove the smell hypothesis I placed only brothers and sisters together in a cage. Results: one female killed its sister, in other cases only one animal gained weight, and in another, all animals stopped growing. Then I tried to eliminate the sense of smell of the aggressors with formalin-soaked cotton balls held in front of their noses for a few seconds. Result: the attacks became even worse.

The situation was indeed dreadful. I had to kill about 20% of the juveniles because isolating them resulted in extremely neurotic behaviour, and the animals could not be handled for experimental purposes.

My next hypothetical explanation was territoriality, possibly occurring extremely early in this species. The territorial behaviour might be elicited by a lack of parental scent-marks in the newly founded Kindergarten cage. If this was true, aggression might be suppressed in a cage already scent-marked by another animal. I therefore introduced an adult male for 24 hours into the Kindergarten cages the day before the juveniles were placed in them. I selected a male which was extremely active in scent-marking and which started to rub its chin and disperse its urine immediately. These scent-marks can be recognised by tupaias for months (van Holst, 1975).

The success of this measure was overwhelming.

The number of despots which had to be isolated immediately dropped from nearly 20% at the end of 1973 to less than 2% in 1975. By now I have no more juvenile despots.

In its natural habitat such a young tupaia would immediately take possession of an area not yet occupied by an adult, scent-marking animal, and would defend it against intruders. This behaviour is probably of selective advantage in a species with such a high reproduction rate.

It is clear now why destroying the sense of smell did not suppress juvenile aggression: to the nose of the little animal the territory remained without an owner. We also have the explanation for the sudden attacks of juvenile animals on their parents which I mentioned earlier. These attacks happened after cage washing, when the cage appeared empty, without an owner, to the young animal.

It is interesting to note that the age at which this complex of territorial behaviour breaks through is obviously not the same in all animals.

I observed that in some families the number of juvenile despots was much higher than in others. Moreover, litter mates mostly behaved quite differently. This indicates that the point of time at which aggression occurs is genetically determined. Individual variation in the ontogenesis of behaviour in tupaias was observed also by Sprankel (personal communication).

In the peaceful animals of the Kindergarten, territorial aggression breaks through at a later age - usually at between 3 and 6 months, sometimes at about one year. This became evident in many young females, which were mated successfully at the age of 3 or 4 months. We observed the typical pair-bonding behaviour, and the female raised one, two or three litters successfully in the presence of the male.

Suddenly, without detectable reason, the female started to attack the male and we had to separate the animals. We then tried to mate the female with another male, sometimes with success, but more often strife occurred again and the females had to be kept single. Possibly they would have tolerated the male in a larger cage.

The conclusions drawn from our observations are: in tupaias the territorial behaviour develops independently of sexual maturity. It may be observed very early. Territorial aggression is elicited in juveniles by a lack of scent-marks of adult tupaias in a territory. It is directed equally towards male and female conspecifics. As long as the animals are within their parents' territory, territorial behaviour is suppressed by their parents' scent-marks.

The age at which territorial aggression can be observed seems determined genetically. It may occur long after fertility has been reached.

We select those animals for the breeding stock in which territorial aggression occurs latest and with the lowest intensity. We found that the longer

TABLE II

The Survival Rate of Newborn Tupaias is Influenced by the Cage Size. Females either devour their babies or do not suckle them if conditions are inadequate

Animals Bred in 18 Months	20 Large Cages 0.5 m^3	15 Small Cages 0.24 m^3
Young born	421	274
Per cage	21	18
Survived	251	119
Per cage	12.5	8
Survival Rate	60%	43%

TABLE III

The Fertility of Female Tupaias, as indicated by Mating Success and Number of Resulting Pregnancies, is Influenced by the Cage Size: 0.125 m^3 for maintenance versus 0.5 m^3 for breeding

	Maintenance Condition	Breeding Condition
Matings	45	168
Copulations	31	162
Success	69%	96%
Pregnancies	20	125
Success related to copulations	65%	77%
Success related to matings	44%	74%

young tupaias are kept in a group or in pairs, the more peaceful will they be later on, when kept single for experimental purposes. In spite of the pronounced territoriality, tupaias obviously need social contact with conspecifics. Animals deprived of such contact struggle, bite and scream when handled. Mating them is unsuccessful; many deprived animals bite their partner's tail-tip which then gets badly sore.

Besides genetic factors, the size of the cage influences the female's tolerance of the male. We keep a certain number of females for experiments during pregnancy in cages 0.24 m^3 in size. Here the females nearly always start fighting during or after their first pregnancy, so that we take the males away as soon as pregnancy is proven by palpation. Fighting animals never raise babies. Therefore, the selection of the most tolerant females for further breeding is only possible in our large cages.

When the females are kept singly, the survival rate of the newborn also depends on the cage size: in our 0.5 m^3 cages it is 60%, in the 0.24 m^3 cages it is 40% (Table II).

A pronounced difference in fertility was observed between females kept in breeding cages and females kept in our maintenance cages, which are 50 cm wide,

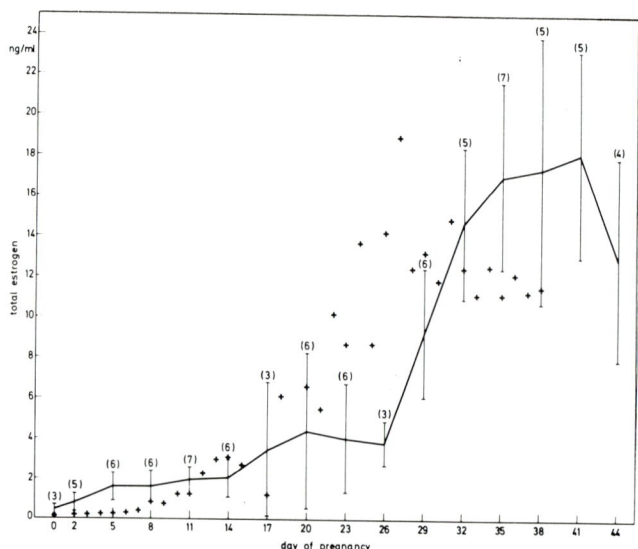

Figure 2. Oestrogen values during pregnancy in Tupaia belangeri (solid line) compared to human values (crosses): in the time scale 1 day in tupaia equals 1 week in the human female.

deep and high (Table III).

In an experiment females in both types of cages, kept there singly for several weeks, were mated once for about 3-6 hours in the morning. Table III shows that receptivity as well as the number of resulting pregnancies from copulations are much lower in the females kept under maintenance conditions.

These mating experiments as well as findings in serial sections of the ovaries of 20 females kept singly, clearly showed that ovulation is induced in tupaias. Large follicles are probably always present in females maintained under favourable conditions.

To rule out the possible exception of spontaneous ovulation following parturition, four females were killed, 12, 24, 24 and 48 hours after parturition without mating. Their ovaries were studied by Kuhn, Göttingen. In no case was spontaneous ovulation observed, and the large follicles became atretic 24 hours after term.

After successful copulation, large corpora lutea develop, which remain preserved until the end of pregnancy. Progesterone levels show an increase during pregnancy in two steps: the first from day 3 on, correlated with the growth of the corpora lutea, the second one beginning around day 20, when the placenta begins to produce progesterone. As in humans, high progesterone levels are reached at the end of pregnancy (Schwaier et al., 1976). The course of total oestrogen levels during pregnancy is also similar to that found in man (Figure 2).

In the non-pregnant tupaia female an oestrous cycle of 10 to 12 days was postulated by Conaway and Sorenson (1966). They had observed copulation cycles in a group of tupaias belonging to different species.

In three of our constantly paired females, which had been good breeders, we ligated the oviducts and ascertained copulation every day by sperm presence in vaginal swabs. In one female the occlusion was not permanent and she became

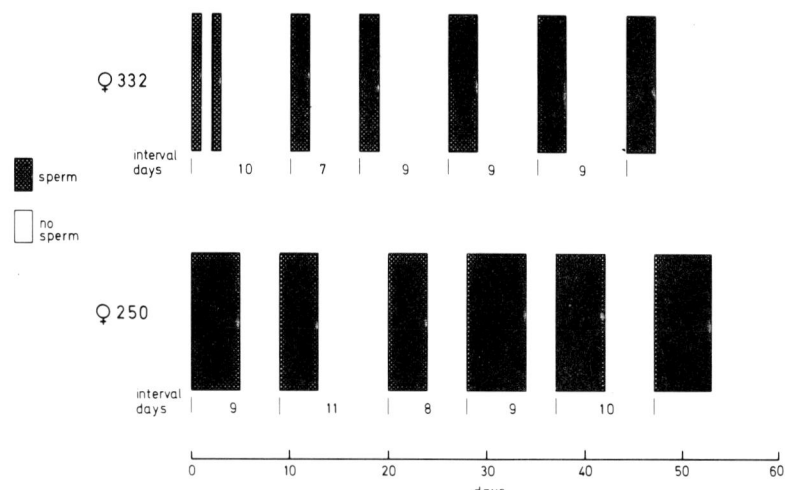

Figure 3. Copulation activity in two pairs of tupaias over seven weeks, as judged from the presence of sperm in the vagina in daily swabs.

pregnant. During pregnancy sperm were present every day except for a short interruption between day 9 and 12. In the other two females we observed that days of copulation alternated regularly with days without copulation (Figure 3). The medium length of one 'period' was 9.3 days.

An unovulatory phase of nine days was also observed following postpartum oestrus. When copulation after parturition did not result in pregnancy, the next bjrth followed with a delay of at least nine days.

These observations, however, do not prove an oestrous cycle of 9 or 19 days. The copulation intervals may simply reflect the life span of the corpus luteum, which suppresses the growth of ovarian follicles. The question of a possible oestrous cycle and its length in tupaias is therefore still open.

In contrast to Conaway and Sorenson we did not observe pseudo-pregnancy or bleeding, except when the females had abortions.

In summary it can be said that breeding tupaias will only be successful if the territorial behaviour of this species is fully taken into account. Moreover, the size and quality of caging determines fertility and breeding success.

In the non-pregnant female there are no parallels to the menstruation cycles of higher primates; however, pregnancy seems to be accompanied by comparable endocrine mechanisms. These studies are continuing at present.

REFERENCES

Conaway, D.H. and Sorenson, M.W. (1966). *In* "Comparative Biology of Reproduction in Mammals" (I.W. Rowlands, ed.). Academic Press, London and New York.
Holst, D. von (1969). *Z. vergl. Physiol.*, **63**, 1-58.
Holst, D. von and Leske, S. (1975). *J. comp. Physiol.*, **103**, 173-188.
Martin, R.D. (1968). *Zeit. Tierpsychol.*, **25**, 409-495 and 505-532.
Schwaier, A. (1975). *In* "Breeding Simians for Developmental Biology" (F.T. Perkins and P.N. O'Donoghue, eds), pp. 141-149, Laboratory Animals, London.
Schwaier, A., Kuhn, H.J. and Hasan, S.H. (1976). *In* "The Laboratory Animal in the Study of Reproduction" (Th. Antikazides, S. Erichson and A. Spiegel, eds), pp. 75-82. G. Fischer Verlag, Stuttgart and New York.
Sprankel, H. (1961). *Zeit. wiss. Zool.*, **165**, 186-220.

THE PRODUCTION OF THE COMMON MARMOSET, *CALLITHRIX JACCHUS*, AS A LABORATORY ANIMAL

W.A. HIDDLESTON

ICI Pharmaceuticals Division,
Alderley Park, UK.

In the mid 1960's it was felt at ICI that there was a need for an additional non-rodent species to the dog to be used in the routine testing and development of new drugs.

Such a species should preferably be a primate which would breed readily in laboratory conditions and would be safe to handle from the aspects of both disease transmission and danger to the experimentalist. What in fact was desired was a primate which would become a laboratory animal and not a zoological specimen used in the laboratory.

Following reports by Kingston and others on the successful use of *Callithrix jacchus* in the laboratory we decided to evaluate this species. We had originally attempted to use this species in 1959 but had abandoned further investigation due to problems of husbandry. The nutritional requirements of this species were now better understood and we decided to re-evaluate it.

Animals were imported from the wild, quarantined and conditioned. There was great variation in quality of imported batches of animals and mortality varied between 10 and 70%. One major factor appeared to be the time of year when animals were trapped - they were always in worst condition during December to February and eventually it was decided not to import in these months. Once delivered the most common problems were enteric infections and parasites. During the conditioning period animals were treated with anthelmintics every 21 days and with the exception of the nematode *Trichospirura leptostoma*, which is present in the pancreas of almost all imported animals, the remaining parasites appear to have been eliminated. One parasite which appeared with increased regularity towards the end of our importing period was the spiny-headed worm, *Prosthenorcis elegans*, which caused high mortality in those batches of animals which were affected. We found them to be resistant to all modern anthelmintics and taenicides but found treatment with oral carbon tetrachloride to be extremely efficacious when administered at the dosage of 0.5 ml/kg bodyweight to starved animals. Although a few animals collapsed following dosage there was no mortality.

Once the animals had been conditioned we subjected them to as many different types of experimental procedure as was practically possible to see where they might be useful in drug development. We were not prepared to establish a breeding colony of animals which would not be used by the experimentalist.

Experimental procedures where the marmoset was used were teratology, where they were found to be susceptible to thalidomide, and toxicology, where the toxicity of compounds was compared in the dog and marmoset. In certain instances where a drug was metabolised differently in the dog and man, the marmoset was found to be preferable because it metabolised the same as in man. In reproductive physiology the marmoset was found to be a useful model while the behavioural scientists also used the marmoset with some success.

The marmoset is susceptible to many human viral conditions and thus may be a suitable model for the study of those diseases.

Simultaneously with the evaluation of the marmoset as an experimental animal we were investigating different methods of husbandry to enable us to arrive at the most efficient means of breeding them on a large scale. We had estimated that if the trials were successful the usage of marmosets would be of the order of 1,000 per year. Any breeding system adopted had to be designed with this scale of usage in mind. Several systems of breeding were tried such as groups of 15 pairs in a gang cage and trios and pairs in cages. Both controlled environments were tried and simple accommodation with heated sleeping quarters and an outside run with no additional humidity. The two groups of 15 pairs did not breed; they lived together for twelve months without any pregnancies going to term although they were frequently seen to be mating. This experiment had to be terminated suddenly as the animals began to fight and had to be caged in pairs to prevent further loss of life.

Despite the fact that they had been running together for 12 months without any pregnancies going to term and that there was no selection when the animals were paired, over 80% of these animals became pregnant within six weeks of being caged. In the case of the trios in cages, only the dominant female became pregnant, the other female acted as a nursemaid for the offspring. When paired up these females bred normally. In the case of the pairs and trios which were allowed to breed in the simple accommodation of a heated sleeping compartment and an outside run the rate of conception and birth was comparable with that of animals in a controlled environment but as the animals went outside in inclement weather many of the babies became chilled and died.

When we established monagamous pairs in a controlled environment we encountered little difficulty in getting them to breed. Twins were the normal outcome of a pregnancy and post partum mating was the rule. Based on this experiment we decided that if we were to establish a breeding unit the only satisfactory method would be in a controlled environment as monogamous pairs.

Since in our initial experiments we were able to wean 3.5 offspring per female per year, we estimated that to produce 1,000 animals for issue each year with adequate replacement breeding stock, 360 breeding pairs would be required.

DESIGN OF UNIT

During our conditioning periods we had experienced some epidemics of infectious disease. In order to lessen the risk we decided to plan our breeding unit as three colonies each with separate staff but common service areas. The accommodation for each colony consists of 10 animal rooms and a diet preparation/record room, the units being serviced by a clean and dirty corridor system, cages being returned via a double-ended cage wash. Each unit can be gassed while the adjacent units are occupied, but not each room separately. Each room is 6 m x 2.5 m and accommodates 40 breeding pairs of marmosets and their offspring until weaning. Temperature is maintained at $24^\circ \pm 1^\circ$ and a relative humidity of $55\% \pm 2\%$ (Figure 1).

Incoming air is filtered to 5 μ, 15 air changes per hour are supplied and the air is not recirculated. The rooms have artificial lighting to 450 lux with a light period of 14 hours; there is no period of subdued lighting. In

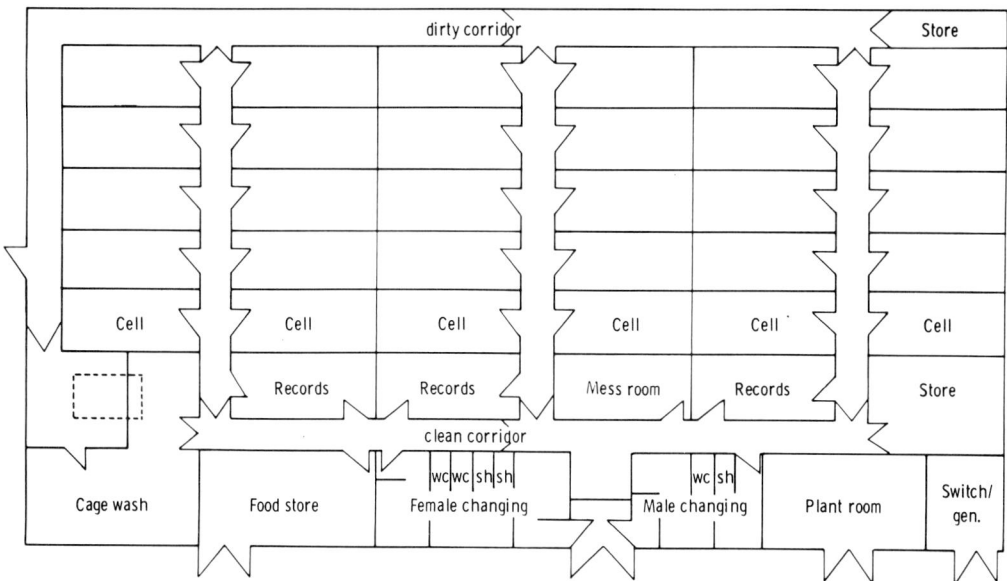

Figure 1.

two rooms in each unit, animals are not caged, an additional door has been provided to make a gang cage. An automatic watering system is provided in all rooms.

Because of problems arising during the period of experimental breeding an electronic alarm system has been installed which will indicate when the environment varies from the design parameters and a standby generator was also installed in the building.

CAGE DESIGN

Each cage is 50 x 50 x 75 cms, with back, sides and top of aluminium sheet, and the front of aluminium mesh. Two doors facilitate handling, and the cages for breeding are provided with shelves 22 x 12 cms while those in the experimental rooms have nest boxes. Each cage has two perches and a perforated aluminium floor with a tray underneath to collect faeces. The cages, each with a drinking nipple, are manufactured in blocks of 4 on castors so that they are easily handled by female staff (Figure 2).

HUSBANDRY

Cages are cleaned daily, although the complete cage is changed and washed fortnightly. We have not experienced any problems we consider attributable to this practice; we feel it is unnecessary even to leave familiar fittings when the cage is changed. Animals are fed daily on a mixture of B P Nutrition Limited expanded primate diet - each pair is allowed 40 g dry diet and 30 g apples and oranges per day; the expanded diet is softened with reconstituted dried 'hi fat' calf milk. The pellet size of the expanded diet proved to be too large, and the consistency too hard for weanling babies to eat. We found the young animals were eating only fruit and milk and evidence of poor weight gain and a wasting syndrome developed, the animals eventually dying of malnutrition. This was corrected by providing the same diet in a more acceptable form and baby marmosets weaned on to a smaller extruded primate

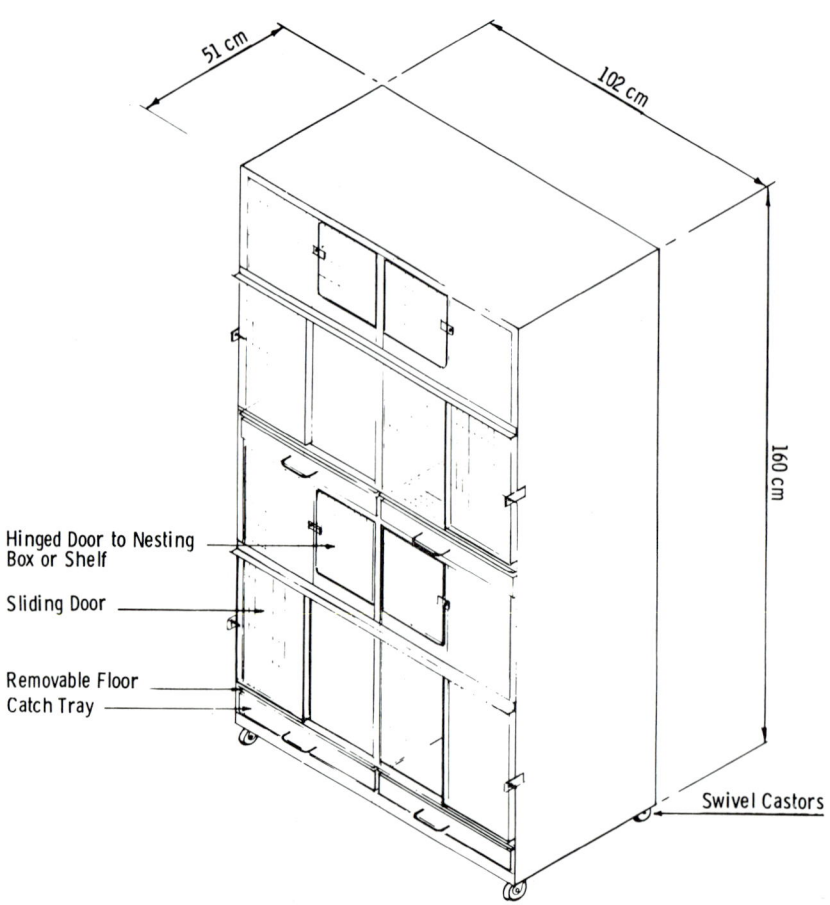

Marmoset 4 Cage Module.
Aluminium Alloy Construction.
Removable Nesting Boxes, Trays
and Base Grills. Fixings For Perch.

Figure 2.

diet pellet. Each time cages are changed vitamin supplements are given in the form of 1,000 iu Vit D_3, and 1 ml Abidec drops. Each month adult animals are treated with anthelmintics using either levamisole (Ketrax syrup) or thibendizole alternately. On only one occasion have we seen endoparasites in a colony-bred animal, and although routine screening is not carried out, all animals are examined at the termination of experiments for the presence of parasites. Any animals which die in the breeding unit are also examined. In jungle animals which have been on site between 5 and 8 years, with the exception of *Trichospirura leptostoma* in the pancreas, parasites are rarely if ever seen.

To establish the breeding unit, 100 pairs of imported animals were selected

TABLE I
Litter Interval Observations

Number of Litters Examined	Mean Litter Interval Parturition→Parturition ± S E	Median Litter Interval	% Litters 140-180 Days
146	178.4 ± 3.7	158 days	72.6

TABLE II
Increasing Incidence of Multiple Births

Year	Number of Parturitions	% Quadruplets	Triplets	Twins	Singles
1	88	0	17.1	71.6	11.3
2	213	0	23.0	73.3	3.7
3	240	.4	24.5	69.2	3.3
4	313	1.9	40.6	51.1	6.4
5 (6 months)	192	.5	41.1	51.5	6.9

and placed in isolation for a period of two years prior to being introduced into the unit. These animals were screened for salmonella and shigella and for the presence of herpes antibodies by the use of *Herpes simplex* antigen. No infected animals were detected and in this period all the animals, with one exception, started to breed, and on completion of the unit, 99 pairs of breeding animals were introduced.

It was found that post-partum mating was the normal practice and that most animals conceived on day 16 post partum; gestation was found to be 142 days with a median litter→litter interval of 158 days. The average age at which the first parturition takes place is 17 months (Table I).

The breeding colony has now been allowed to expand to 320 pairs and it is planned to allow further expansion to 360 pairs. Some of the imported animals which were mature on arrival in 1968 continue to breed well and rear their offspring. When we were planning the colony we estimated we would get possibly 4 or 5 years production from one female; this estimate would now appear to be conservative and the useful breeding life may be as much as double this.

One change which has been noticeable in the progress of the colony has been the increasing incidence of multiple births (Table II).

I attribute the increase in multiple births to the higher plane of nutrition of all the animals and to the elimination of parasites from the imported animals. Since in the vast majority of cases the female marmoset will only rear twins, we investigated the cost-effectiveness of hand rearing the surplus babies, but concluded that it was not a viable proposition bearing in mind the extra staff required, although technically it was perfectly feasible.

Since a small proportion of animals do manage to rear all three babies, it may be possible in the future to select replacement breeding stock for maternal and lactating ability, thereby establishing a strain in which the number of triplets reared is high. It is our practice, if all three babies are alive after 7 days, to feed them twice daily with a mixture of 50% SMA and 50% Complan from a syringe, and with this help the parents manage to rear all three babies.

In the 1,046 pregnancies which have so far gone to full term, 2,351 babies have been born and 1,640 animals have been weaned. Abortions are uncommon, only 15 having been observed over this period, and as animals are not routinely palpated, we have no information on foetal resorption. The abortions have normally been associated with a change in the environment, either transferring the animals to another group or the presence of contractors working with power tools in the building. In this period 87 stillbirths have been observed. Of the causes of death in unweaned marmosets, 45% have been surplus triplets. The next most common cause of death was lack of care of first litters by inexperienced mothers. In subsequent litters two babies were reared. As the number of experienced mothers in the colony increases, I anticipate the preweaning mortality rate will decrease.

Other causes of losses have been agalactia where we have been experimenting with diets without fruit, and mastitis in aged females. There have been low incidences of pneumonia and enteritis in unweaned babies. Some deaths have occurred as a result of improper presentation of the diet.

As weaning approaches at between 14 and 16 weeks the special diet is introduced with the parents food and the unweaned babies become accustomed to eating it. Prior to weaning, babies are vaccinated with an *E. coli* vaccine. In the early stages of development of the colony animals were found dead some 14-21 days post weaning and within a few hours of appearing normal; an *E. coli* enterotoxin was isolated. A commercially available piglet enteritis vaccine was found to contain the appropriate strains of *E. coli* and has since been used successfully as a prophylactic in our marmosets.

On weaning, the animals are maintained as pairs for two to three weeks until they become established and are seen to be eating adequate amounts of diet. They are then turned loose in a gang cage in groups of up to 60 where they will remain until they become mature. During this period the staff are encouraged to handle the animals so that when they are again caged for experimental use they are accustomed to human contact and to being handled. About 4 weeks prior to being issued the animals are caged in pairs of the same sex. At this stage it is easy to pair females and the pairs seldom have to be rearranged. Once a female has become accustomed to being with a member of the opposite sex, it becomes almost impossible to cage with another female as they fight viciously. There is little problem in pairing males.

Despite the fact that the young marmosets are weaned and removed from their parents' cage prior to the next litter being born they have few problems in rearing their own offspring once they become mature and start to breed. There is a higher preweaning loss among first litter marmosets as with other species.

Patterns of normal growth (Figure 3) have been established for our animals following weaning and we are now in a position to issue animals on an age/weight basis similar to dogs or other laboratory animals. Blood samples are taken from the femoral vein with the animal restrained on an apparatus. Up to 5 x 1 ml samples have been taken in one day. The normal parameters we obtained are shown in Tables III and IV.

DURATION OF PREGNANCY

One of the problems which required to be overcome was the accurate establishment of the time of successful mating. Using a combination of the history

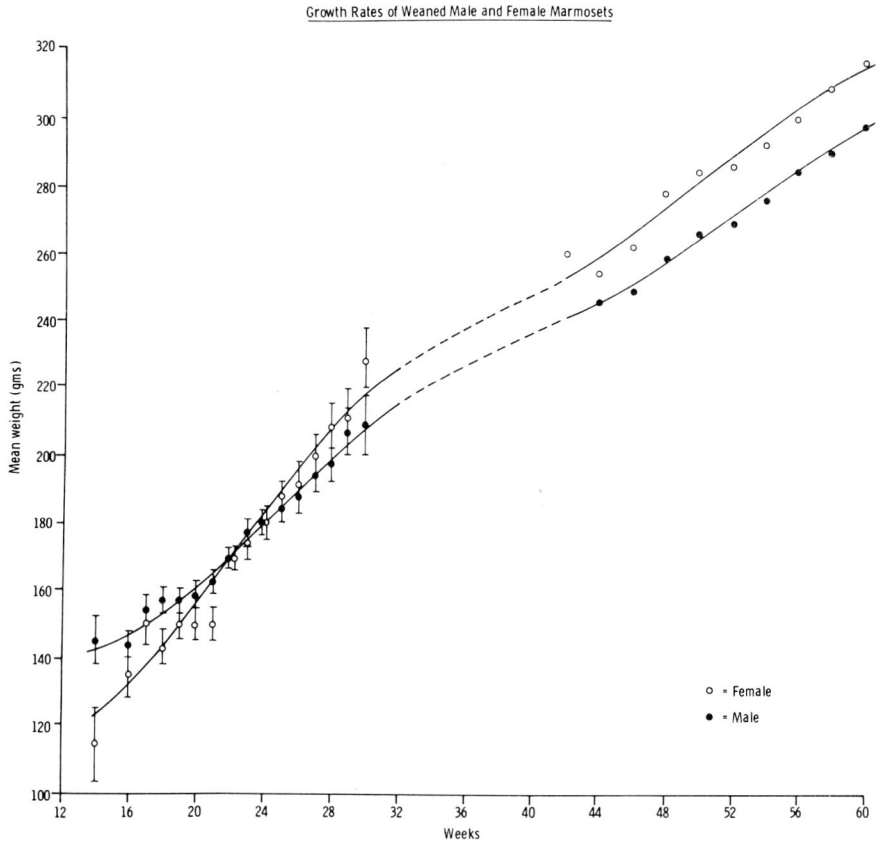

Figure 3. Growth rates of weaned male and female marmosets.

of the animal and palpation, it is now possible to estimate the stage of pregnancy to ± 24 hours sufficiently early to enable pregnant animals to be dosed during the period of organogenesis. Measurements of the pregnant uterus with calipers will indicate the age of the pregnancy.

NUTRITIONAL PROBLEMS

In our early experiences with marmosets we discovered a high incidence of fatty infiltration of the liver in animals which underwent post mortem examination for a variety of reasons. At this time malt bread was being fed as a high proportion of the diet. The malt bread was removed from the diet and the condition disappeared. Experiments were carried out to determine whether marmosets could be maintained successfully but although they bred and produced live offspring they did not lactate sufficiently to rear their offspring. If fruit was added the numbers of offspring reared increased dramatically. We also inadvertently discovered a Vitamin C deficiency during some of the experiments, with diets which had been stored too long. This presented as:

TABLE III
Haematology in Marmosets

Test	Units	Range Seen
Haemoglobin	g/dl	14.0-17.5
Packed Cell Vol	%	44-52
Total WBC	/cu mm	5,200-17,500
Neutrophils	/cu mm	1,320-6,300
Lymphocytes	/cu mm	4,092-11,285
Platelets	/cu mm	220,000-500,000

TABLE IV
Clinical Chemistry: Normal Values in Marmosets

Tests and Units	Males			Females		
	No.	Mean	Range: 95% Confidence Limits	No.	Mean	Range: 95% Confidence Limits
Sodium (m mol/L)	21	1/9.1	119-162	27	153.7	135-165
Potassium (m mol/L)	36	4.6	1.7-7.1	39	5.0	3.0-7.4
Calcium (m mol/L)	18	2.57	2.05-2.95	17	2.72	2.35-3.35
Inorganic Phosphorus (m mol/L)	20	2.03	1.00-3.16	19	2.22	1.38-3.23
Glucose (m mol/L)	51	6.9	1.3-22.9	47	7.7	2.9-28.1
Urea (m mol/L)	49	11.6	3.6-22.8	40	11.6	5.7-23.1
Total Protein (g/L)	22	70	56-82	22	73	60-82
Albumin (g/L)	52	41	22-52	51	42	34-60
Total Bilirubin (mol/L)	15	14	7-21	13	10	3-21
Alk Phos U/L	21	203.7	87-332	18	195.2	73-335
AST U/L	47	129.6	43-250	53	106.1	50-186
ALT U/L	47	9.4	3-30	42	8.0	3-24
ICDN U/L	11	64.9	34-93	12	61.8	26-101
CPK U/L	20	284.9	119-891	22	254.6	40-840

1. Animals losing weight, and anorexia.
2. The animals became stiff and unwilling to move. They sat on the cage floor rather than on perches.
3. Petechial haemorrhage into the eyelids.

The addition of ascorbic acid to the drinking water produced a dramatic improvement in the condition and diet intake of these animals and there was an eventual recovery.

Osteomalacia or rickets has been seen in unweaned and weaned animals, and we now give 1,000 iu Vitamin D_3 fortnightly as a prophylactia.

REPRODUCTIVE PROBLEMS

Marmosets on arrival from the wild were paired on the basis of the first male unpacked being put with the first female. It was very rare that any trouble was experienced with incompatibility between the animals. In less than 3% of the animals was it necessary to rearrange the pairings, somewhere in the order of 1,000 pairs having been made up in this fashion. This process has been extended to our home bred animals and apart from checking that we are not pairing brother and sister, no other selection for compatability is carried out. However, animals are selected for size and for mothering qualities. At present the entire colony is being surveyed for chromosomal abnormalities.

Should one of a pair become discarded for any reason, or die, we do not experience any difficulty in pairing them with another mate.

We have recently experienced a condition which can best be described as imperforate vaginas in a number of F1 and F2 animals, mainly F2; the external female genitalia fail to develop completely. This condition may be hormonal or dietetic in origin, or both, other theories are that it may be related to the occurrence of twins of different sexes in the same litter, or related to the offspring of a female with an XY chromosome. We have observed such a female in our colony. In any case, I am uncertain of the aetiology. Records are being kept of the incidence to try to find out more about it.

Dystocia has been uncommon; only 11 cases out of 1,046 pregnancies.

COSTS

Since each institution has its own system of costing its animal production, I feel there is little value in quoting actual costs of production of marmosets. However, since most institutes cost all their animals in the same way and can compare relative costs, I have prepared a table of comparative costs using our own system.

One marmoset at age of issue 10-12 months, costs:
10% less than 1 dog at 10 months of age;
Equivalent to 3 mini pigs at 15 kg in weight;
Equivalent to 40 guinea pigs;
Equivalent to 84 rats;
Equivalent to 200 mice.

I feel we have now taken the common marmoset to the point where it can be considered as a laboratory animal. It can be bred in the laboratory in large numbers. Its growth and development can be predicted and under our systems of husbandry we are aware of many of its normal physiological parameters - it is no longer a zoological species used in the laboratory.

BREEDING MARMOSETS FOR MEDICAL RESEARCH

S.F. LUNN AND J.P. HEARN

*MRC Unit of Reproductive Biology,
2 Forrest Road, Edinburgh EH1 2QW UK.*

The present communication discusses various aspects of the breeding of the common marmoset, *Callithrix jacchus*.

COLONY GROWTH

During the period January 1973 to July 1976 inclusive, 156 wild-caught marmosets were obtained, a further 38 animals being supplied from established colonies. Throughout this time 275 marmosets were born, giving a theoretical total of 469 marmosets; however, the actual colony size was 265. Only 74 wild-caught marmosets have survived to the present time. Fifty-six deaths occurred within six weeks of arrival (equivalent to the average conditioning period of the dealer), and this figure had risen to 67 by the end of the statutory six month rabies quarantine period. The reasons for these relatively high losses included the marked degree of dehydration and emaciation seen on the arrival of the animals, the stresses associated with their capture and transport (particularly in very young or old marmosets), and parasitic infestation of the intestine, exacerbated by loss of condition and change of diet. In addition, a viral infection, tentatively identified as rubella, was responsible for many deaths in one wild-caught intake.

Two-thirds of the animals obtained from established colonies are still alive; the breakdown of figures for the 275 animals born in the colony is given later on in this paper.

The growth of the colony was less than maximal since many of the animals were involved in research projects which affected their reproductive potential. These included orchidectomy, vasectomy and ovariectomy, with or without hysterectomy. In addition, marmosets were immunized actively and/or passively against the β-subunit of human chorionic gonadotrophin or synthetic luteinizing hormone-releasing hormone.

INTERBIRTH INTERVAL

The mean interval between successive births in 36 animals (n = 69) was 247.5 ± 91.2 (SD) days, values ranging from 151-649 days. No attempt was made to shorten this period by hand-rearing the infants. As part of the routine colony management, all post-pubertal female marmosets undergo transabdominal uterine palpation for the detection of pregnancy (Hearn and Lunn, 1975). This procedure also detects abortion, a term which is being used here to include both foetal resorption and miscarriage, whether spontaneous or induced. Elimination of interbirth intervals during which one or more abortions occurred caused the mean interval to decrease to 209.5 ± 52 days (n = 48 in 27 animals: range = 151-384 days).

It is difficult to assess foetal wastage due to premature termination of pregnancy, since products of abortion are generally eaten and uterine palpation is not the most sensitive index of its occurrence; however, it may be high, Hampton et al. (1971) reporting that 20.9% of pregnancies in their *C. jacchus* colony ended in abortion.

POST-PARTUM MATING

Daily vaginal lavage was performed in 24 animals from the day of parturition until sperm were found to be present. Mating occurred on day 8.7 ± 8.9 days (range = 2-47 days) and was in agreement with an earlier study (Hearn and Lunn, 1975). The large SD was due to the fact that in two of the twenty four animals, sperm were first observed at 21 and 47 days post-partum.

Gestation length in the marmoset is approximately 21 weeks (Hearn and Lunn, 1975) and lactation can continue for 90-100 days post-partum. Since post-partum mating usually occurs within 1-2 weeks, and the interbirth interval can be as short as 151 days, it can be seen that early post-partum mating can result in conception, and that this is not necessarily inhibited by lactation.

COLONY-BRED ANIMALS

Of the 275 marmosets born in the colony, 29 (10.5%) were stillborn, 80 (29.1%) died and 166 (60.4%) are currently alive, i.e. of 246 live born animals, 67.5% have survived.

Of the 80 deaths, five could be attributed to accident. Of the remaining 75, > 50% had died within five days of age, this value rising to 90% by 50 days of age.

The incidence of stillbirths was less than the 26.8% recorded by Phillips (1976), but was still considered to be relatively high. Stillbirths were not related to increased number of young at birth, 15 being associated with triplet and one with quadruplet births. The remaining 13 stillbirths occurred with single and twin deliveries.

Early post-natal deaths occurred relatively frequently, and this was particularly so with triplet deliveries, where in no case were the parents able to rear all three successfully.

SEX RATIO

The sex of 6 of the 275 animals bred in the colony could not be ascertained because of partial cannibalization. Of the remainder, a sex ratio (M:F) of 140:129 was recorded. A X^2 test reveals that this does not differ from an expected 1:1 ratio. Data from nine marmoset colonies (*Callithrix* and *Saguinus* sp) have been collected by Ford and Evans (1977). On 2,741 animals in which the sex was determined, there was a statistically significant preponderance of males, and re-analysis of their data to include the present values does not influence this finding.

INCIDENCE OF TWINNING

Of 127 births there were 15 single young (11.8% of total births), 78 sets of twins (61.4%), 32 sets of triplets (25.2%) and two sets of quadruplets

(1.6%). For the twin births, a ratio for M/M:M/F:F/F of 1:1.6:0.7 was recorded. This approaches a frequency of 1:2:1 expected for dizygotic twinning and is in agreement with reports from others (see *int. al.* Hampton et al., 1971).

In conclusion, it would appear that a self-replacing, economically and scientifically acceptable colony of non-human primates can be established readily, if the animal chosen is the common marmoset.

REFERENCES

Ford, C.E. and Evans E.P. (1977). *J. Reprod. Fert.*, 49, 25-33.
Hampton, J.K. Jr., Hampton, S.H. and Levy, B.M. (1971). Reproductive physiology and pregnancy in marmosets. *In* "Medical Primatology 1970" (E.I. Goldsmith and J. Moor-Jankowski, eds), pp. 527-535. Karger, Basel.
Hearn, J.P. and Lunn, S.F. (1975). *In* "Breeding Simians for Developmental Biology" (F.T. Perkins and P.N. O'Donoghue, eds), pp. 191-202. Laboratory Animals, London.

AN ARTIFICIAL MILK FOR HAND-REARING SPECIFIED PATHOGEN FREE MARMOSETS, *CALLITHRIX JACCHUS*, AND THE GROWTH OF ANIMALS ON THE PREPARATION

J.A. TURTON, K.R. HOBBS, D.J. FORD, J. BLEBY AND B.M. HALL*

Medical Research Council, Laboratory Animals Centre, Woodmansterne Road, Carshalton, Surrey.
The Nuffield Institute of Comparative Medicine, Zoological Society of London, Regent's Park, London.

INTRODUCTION

A breeding colony of marmosets (*Callithrix jacchus*) was established at the Laboratory Animals Centre (LAC) in 1969 and a project initiated to derive specified pathogen free (SPF) animals by hysterotomy and hand-rearing (Hobbs et al., 1977). The project was divided into three parts: first, the formulation of an artificial marmoset milk; second, the hand-rearing of naturally born infants under conventional (non-SPF) conditions; and last, the hand-rearing of hysterotomy-derived young using aseptic procedures, contamination of the infants with a bacterial flora, and the transfer of the young into the SPF unit for the completion of the rearing programme.

ARTIFICIAL MILK

To formulate an artificial marmoset milk, lactating females were milked and the milk analysed (Hobbs et al., 1977; Turton et al., 1977). The results are summarized in Table I. A human milk substitute formed the basis of the

TABLE I

Analysis of Human Milk, a Human Milk Substitute, 2 Artificial Marmoset Milks, and Natural Marmoset Milk

	Crude Protein (g/100 ml)	Lactose (g/100 ml)	Total Lipids (g/100 ml)	Energy Level (kcal/100 ml)
Human milk[a]	1.5	6.8	4.0	68
Human milk substitute[b]	1.5	7.2	3.6	65
Aberystwyth marmoset milk formulation[c]	6.8	5.7	3.7	83
LAC marmoset milk formulation	3.3	5.1	7.6	102
Marmoset milk	3.6	7.5	7.7	114

[a] Data from Davidson et al. (1975).
[b] Gold Cap S-M-A Ready-to-Feed milk (John Wyeth and Brother Ltd.); data from product information literature.
[c] As described by Stevenson (1976).

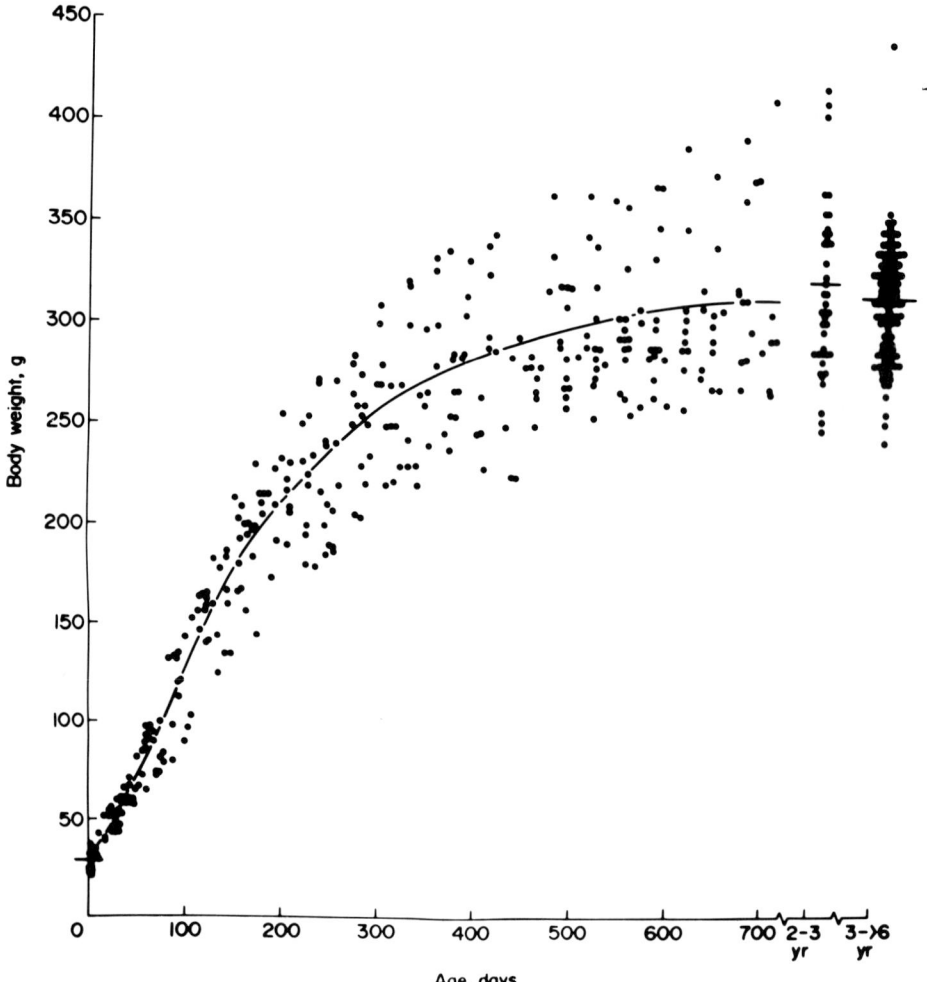

Figure 1. Increase in body weight of male marmosets. Each point is the weight of a single animal; both singletons and twins are included. The curve was drawn in by eye. Mean body weights (shown by horizontal lines): at birth, 29.3 g (sd 4.1 g); at 2 to 3 years, 324.2 g (sd 44.9 g); at > 3 years, 317.4 g (sd 28.7 g).

artificial marmoset milk, and an analysis of this preparation (Table I) indicated that the levels of protein and total lipids should be increased. Accordingly, the 'LAC formulation' was prepared: 76 ml Gold Cap S-M-A Ready-to-Feed milk (John Wyeth and Brother Ltd.); 24 ml Pure Dairy Sterilised Cream (Nestlé's Cream; The Nestlé Co. Ltd.); 2 g calcium caseinate (Casilan; Glaxo-Farley Foods, Ltd.). Table I gives the analysis of this milk, and details are also given of the preparation used successfully by Stevenson (1976) to hand-rear *Callithrix jacchus*.

HAND-REARING CONVENTIONAL MARMOSETS

A convenient way to assess the growth of hand-reared primates is to compare the body weight increases with those of mother-reared counterparts. The

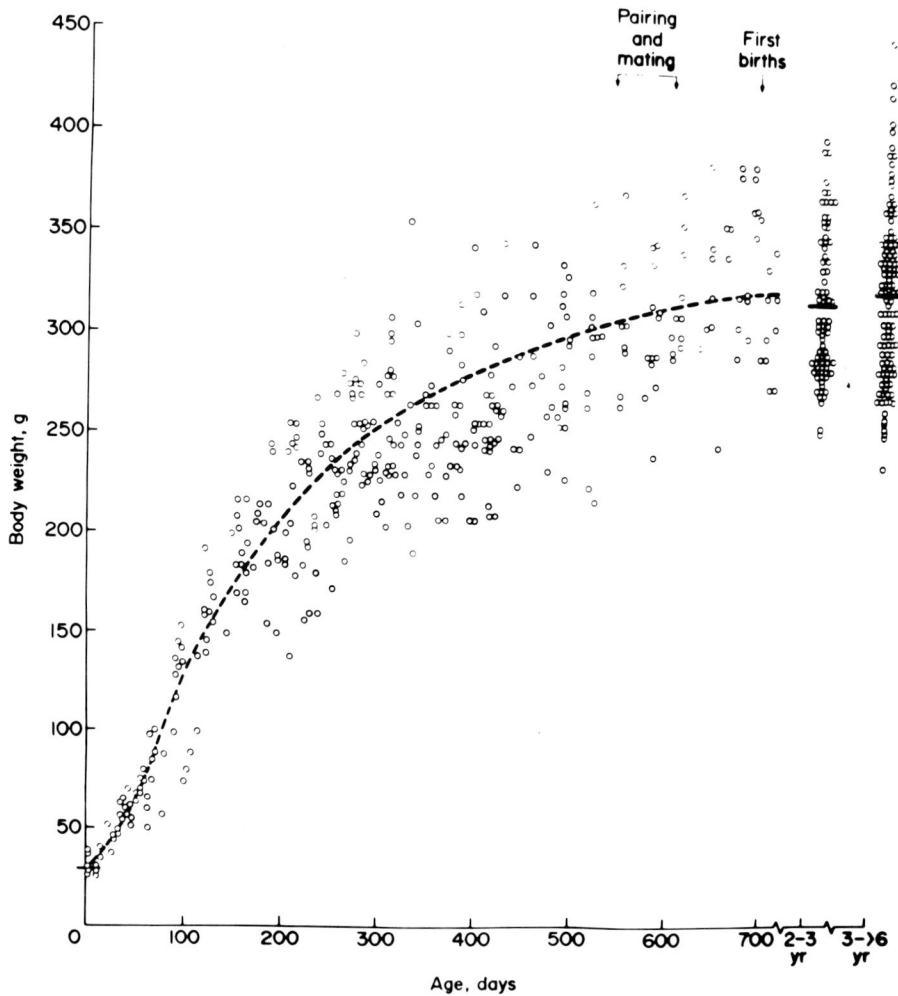

Figure 2. Increase in body weight of female marmosets. Each point is the weight of a single animal; singletons and twins, and pregnant and lactating animals are included. The curve was drawn in by eye. Mean body weights (shown by horizontal lines): at birth, 28.7 g (sd 4.0 g); at 2 to 3 years, 318.1 g (sd 36.5 g); at > 3 years, 323.7 g (sd 42.7 g).

increases in body weight of male and female marmosets of the conventional LAC colony are presented in Figures 1 and 2. In general, the body weight increases compared with those reported by Epple (1970), Hearn and Lunn (1975) and Hiddleston (personal communication). There were no clear differences in the growth of male and female animals, confirming the findings of Hearn and Lunn (1975).

The LAC milk formulation was first employed as a support feed, that is, it was given to infants being suckled naturally. In the first instance a male of 14 days was fed small amounts (1 to 3 ml) of the preparation each day. At about 30 days of age a human baby cereal (Farex or Farlene; Glaxo-Farley Foods Ltd.) was mixed with the artificial milk. The increase in body weight of this infant is illustrated in Figure 3. In this way the techniques employed in

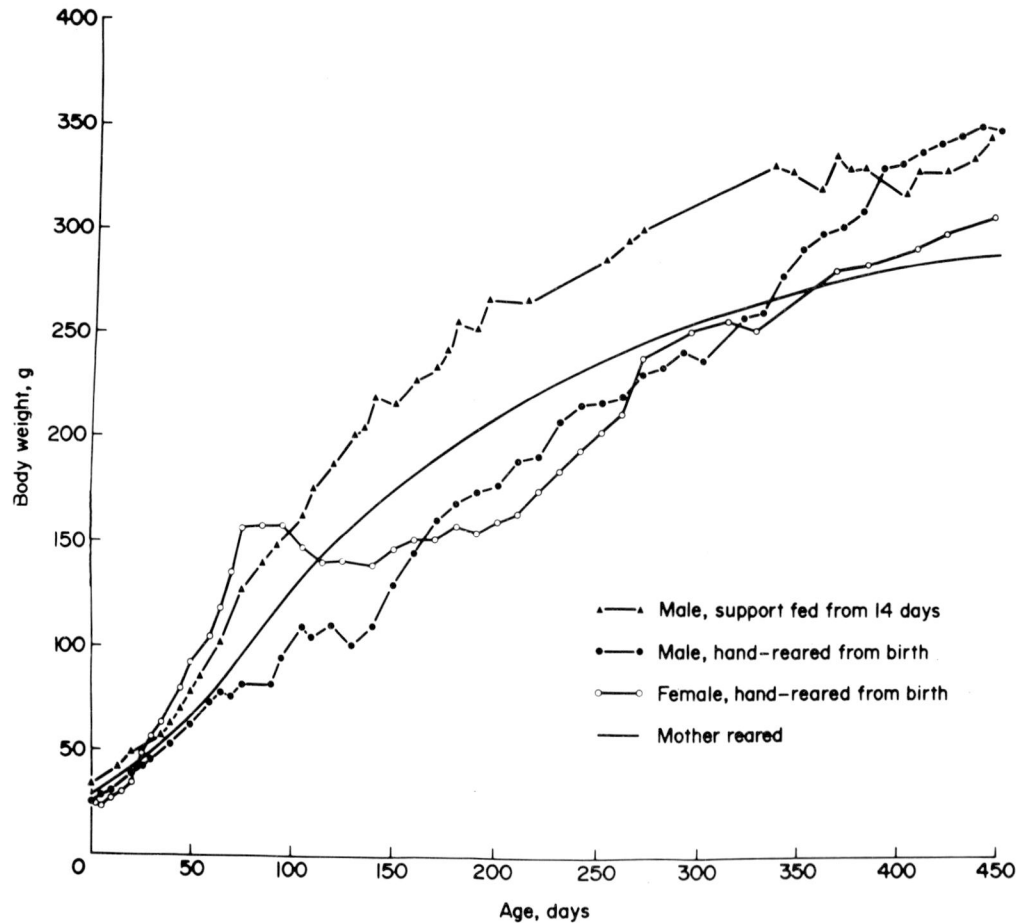

Figure 3. Increase in body weight of naturally-born, hand-reared, non-SPF infants fed the LAC milk formulation from birth, or support fed from day 14. The increase in body weight of mother-reared animals from the conventional colony is included for comparison (this curve is the mean of those presented in Figures 1 and 2).

hand-rearing were practised safely, and the artificial milk and baby cereals evaluated.

The next stage was the hand-rearing of infants born naturally, but removed from their parents at birth. Details of the techniques employed have been reported by Hobbs et al. (1977). For the first 24 hours neonates were fed every 2 hours with the artificial milk diluted with a 5% glucose solution, and thereafter the milk alone was given every 2 hours. After the second day, feeding from midnight to 0800 hours was stopped. Small amounts of baby cereal were added to the milk at day 10, and the quantity of cereal gradually increased. A tinned infant beef and liver soup was included in the diet at approximately 30 days, and small quantities of egg at about day 60. Fruit was added to the feeding regime at approximately 90 days, and at this time the infants were introduced to the compounded primate diet (Mazuri; BP Nutrition (UK) Ltd.) being fed to the main colony. The dietary regimes were flexible in content and volume, allowing for individual variations. The body weight increases of 2 animals reared in this fashion are illustrated in Figure 3. In general,

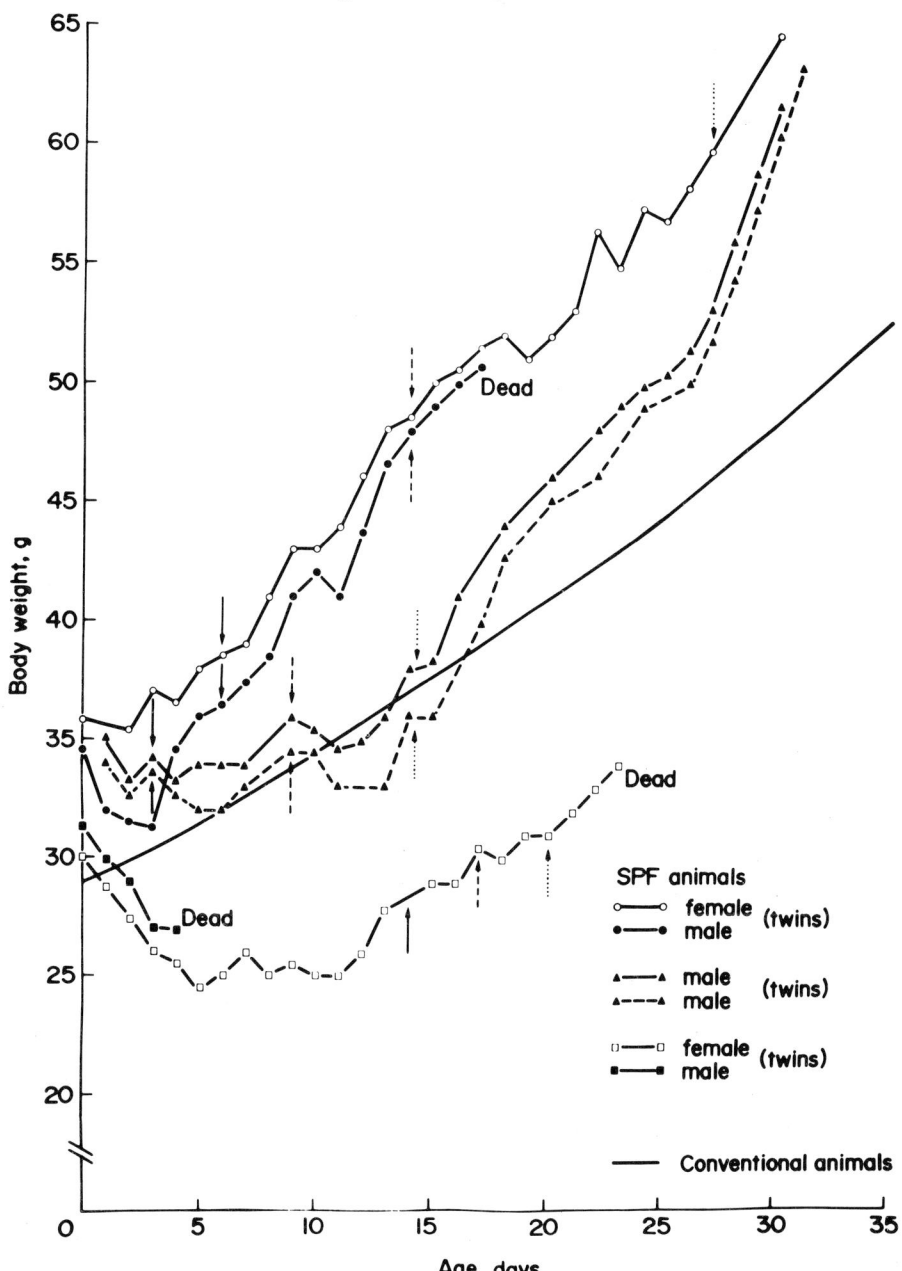

Figure 4. *Increase in body weight of hysterotomy-derived, hand-reared SPF infants fed the LAC milk formulation. The increase in body weight of mother-reared young from the conventional colony is included for comparison.* ↓ *, contamination with Lactobacillus acidophilus;* ↓ *, contamination with a mixed bacterial inoculum;* ↑ *, transfer into the SPF unit.*

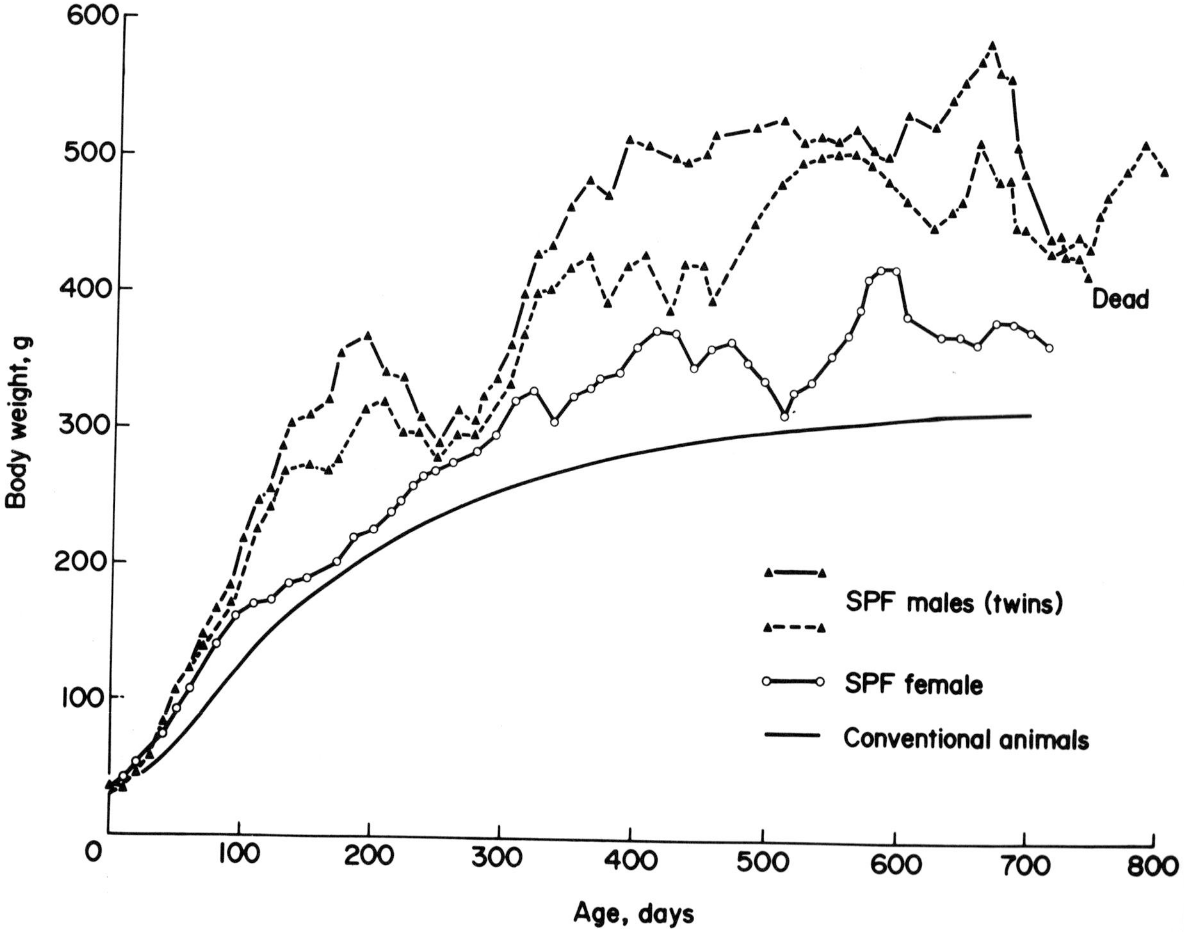

Figure 5. Increase in body weight of hysterotomy-derived, hand-reared SPF animals from birth to approximately 750 days. The increase in body weight of mother-reared young from the conventional colony is included for comparison.

weight gains compared favourably with those of mother-reared infants. It is seen in Figure 3, however, that the female failed to increase in weight from approximately 75 to 150 days. During this period efforts were being made, with some difficulty, to change the diet gradually over to the compounded primate food; these attempts were eventually successful.

From these studies it was established that the artificial milk and the dietary additions were satisfactory, for the infants grew well, paired with naturally reared animals, bred and raised their young successfully.

HAND-REARING SPF MARMOSETS

The final step was the derivation of SPF marmosets. Employing sterile hysterotomy techniques, young were removed at term into a sterile surgical isolator and on resuscitation were passed into a rearing isolator. Feeding procedures were, in the main, as outlined previously. Generally, at 3 to 6 days post-partum the young were contaminated orally with *Lactobacillus acidophilus*, and a van der Waaij mixed bacterial inoculum (van der Waaij et al., 1971) was given per rectum at 9 to 17 days. However in one case an inoculum derived from a faecal preparation of marmosets already in the SPF building was used. Infants were passed into the SPF unit between 14 and 27 days after birth. Increases in weight of hysterotomy-derived SPF young are shown in Figure 4.

Over a 4 year period a total of 17 hysterotomies were performed and 39 animals delivered. Many died at 4 to 15 days after delivery. In several cases bacteriological investigations conducted at post-mortem examination revealed *Escherichia coli* or *Pseudomonas* bacteraemias. Four animals were transferred into the SPF unit, one dying 3 days later of an *E. coli* bacteraemia. The remaining 3 infants, 2 males and 1 female, thrived (Figure 5). However, at 107 weeks of age the dominant male died of a haemorrhagic enteritis associated with an *E. coli* bacteraemia. At present, the remaining pair are a 3 year 10 month old male and a 3 year 7 month old female, but these animals have not bred.

CONCLUSIONS

It would appear that the milk formulation and the later dietary additions were satisfactory when judged by the criterion of body weight increase. However, at present, the most important problem in the derivation of SPF marmosets is the bacterial contamination of the neonate. Although this difficulty has not been overcome, studies are now being carried out to gain a better understanding of the bacteriology of the marmoset gut. It is hoped that this information, in conjunction with the use of vaccines and antibiotic chemotherapy, will facilitate the establishment of a functional SPF breeding colony.

REFERENCES

Davidson, S., Passmore, R., Brock, J.F. and Truswell, A.S. (1975). "Human Nutrition and Dietetics". Churchill Livingstone, Edinburgh.
Epple, G. (1970). *Folia primat.*, 12, 56-76.
Hearn, J.P. and Lunn, S.F. (1975). *In* "Breeding Simians for Developmental Biology" (F.T. Perkins and P.N. O'Donoghue, eds), pp. 191-204. Laboratory Animals, London.
Hobbs, K.R., Clough, G. and Bleby, J. (1977). *Lab. Anim.*, 11, 29-34.
Stevenson, M.F. (1976). *In* "International Zoo Yearbook" (P.J.S. Olney, ed.), Vol. 16, pp. 110-116. Zoological Society of London.
Turton, J.A., Ford, D.J., Bleby, J., Hall, B.M. and Whiting, R. (1977). *Folia primat.*, In Press.
Waaij, D. van der, Vries, J.M. de and Lekkerkerk, J.E.C. (1971). *J. Hyg. Camb.*, 69, 405-411.

REPRODUCTION IN THE SQUIRREL MONKEY (*SAIMIRI SCIUREUS*)

W. RICHARD DUKELOW

*Endocrine Research Unit, Michigan State University,
East Lansing, Michigan USA.*

In 1973 over 55,000 non-human primates were imported into the United States of which 26% were of New World species. Of these, 5,600 were *Saimiri sciureus* and this is the second most commonly used non-human primate for research. Also in 1973, 185 births of *S. sciureus* were reported in the United States, representing about 3% of the animals needed for research. Even this estimate is high since many of the births were in zoos or commercial primate displays and thus unavailable for research. In 1976 the United States Wildlife Service proposed laws that would place *S. sciureus* on the list of threatened species, further restricting the availability of this research animal used in various research endeavours ranging from cancer and arteriosclerosis research to contraceptive and teratological testing. Thus, with decreasing availability of this valuable research animal from the wild and its increasing usage in research laboratories, the necessity for domestic breeding is evident.

Fortunately, a background of strong basic reproductive information exists for *S. sciureus*. The work of Bennett (1967a, b) clearly demonstrated the feasibility of ovulation induction, semen collection, and artificial insemination in this species. In 1970 Dukelow reported on studies designed to control ovulation to single or double ovulations as opposed to the super-ovulation regimes of Bennett. In 1973 Harrison and Dukelow reported a seasonal responsiveness to this regime and, subsequently, a means of overcoming the low seasonal response during the summer months (Kuehl and Dukelow, 1975a, b; Jarosz and Dukelow, 1976).

Several groups have been successful with natural breeding of *S. sciureus*. DuMond (1968) has shown considerable success in a seminatural environment in Florida. Similarly, Clarkson et al. (1968) have had good breeding results with indoor-outdoor caging in North Carolina. Successful indoor breeding has been reported by Bantin (1966), Rosenblum and Cooper (1968), and Hupp (1972). Jarosz and Dukelow (1976) presented preliminary results with outdoor breeding after ovulation induction and have verified the fertility of ova obtained by induction with both natural breeding and *in vitro* fertilization (Kuehl and Dukelow, 1975; Dukelow and Kuehl, 1975).

The present paper reports the ovulatory response to gonadotropic induction from 1971 to date including three periods of induction during the summer or Michigan anovulatory season. The effects of these ovulatory regimes on mating behaviour and vaginal cytology during the ovulatory and anovulatory seasons are also presented.

MATERIALS AND METHODS

All *S. sciureus* used were of either Columbian or Bolivian origin (Tarpon

TABLE I

The Effectiveness of Various Ovulation Induction Regimes

Regime	Number of Animals Ovulating/ Total Animals	Percent
5 days progesterone; 4 days FSH (1 mg); HCG (500 iu)		
Jan-Feb	7/24	29.2
Mar-Apr	11/18	61.1
May-Jun	12/36	33.3
Jul-Sep	9/72	12.5
Oct-Dec	24/58	41.4
Summer Season (July-September)		
5 days FSH (1 mg); HCG (500 iu)	4/8	50.0
5 days FSH (1 mg); HCG (1000 iu)	4/8	50.0
5 days FSH (1 mg); HCG (1500 iu)	4/8	50.0
4 days FSH (2 mg); HCG (500 iu)	3/8	37.5
4 days FSH (2 mg); HCG (1000 iu)	5/8	62.5
4 days FSH (2 mg); HCG (1500 iu)	0/8	0
Sequential: 4 days Estradiol (20 µg); 5 days progesterone		
4 days FSH (1 mg); HCG (500 iu)	7/34	20.6

Springs Zoo, Inc., Tarpon Springs, Florida and Primate Imports, Port Washington, New York respectively). Until May 1975 all animals were maintained in temperature-controlled (20-21°), light-controlled (12-12 h light-dark schedule) quarters in double unit, modular, stainless steel cages with three to five animals per cage. From May 1975 until November 1975 the experimental animals were housed in outdoor cages (12 females and 2 males per cage) with access to an enclosed shelter box. In 1976 the animals were held in the same area from mid-April until November. The animals were fed twice daily with a commercial pelleted feed and had access to water *ad libitum*. Over the years of captivity these *S. sciureus* were used for a variety of experiments, which have been summarized by Dukelow (1975). These involved laparoscopic observation of the ovaries and recovery of ova for use in related experiments. The standard ovulation induction regime consisted of a pretreatment of four days of 5 mg of progesterone daily to mimic the luteal phase of the cycle, followed by four days of 1 mg follicle-stimulating hormone (FSH-P, Armour and Co., Chicago) and, on the fourth day, 500 iu of human chorionic gonadotropin (HCG) (Dukelow, 1970). During the anovulatory season it is necessary to increase the level of FSH to obtain a response. This is done either by increasing the level of FSH to 2 mg daily or by giving 1 mg FSH for five days instead of four (Kuehl and Dukelow, 1975). Studies have also been conducted using a sequential oestrogen-progesterone pretreatment prior to the four day FSH-HCG regime (Kuehl and Dukelow, 1975; Jarosz and Dukelow, 1976). Detection of ovulation and diagnosis of pregnancy were done laparoscopically as we have described earlier (Dukelow, Jarosz, Jewett and Harrison, 1971). During the spring and summer

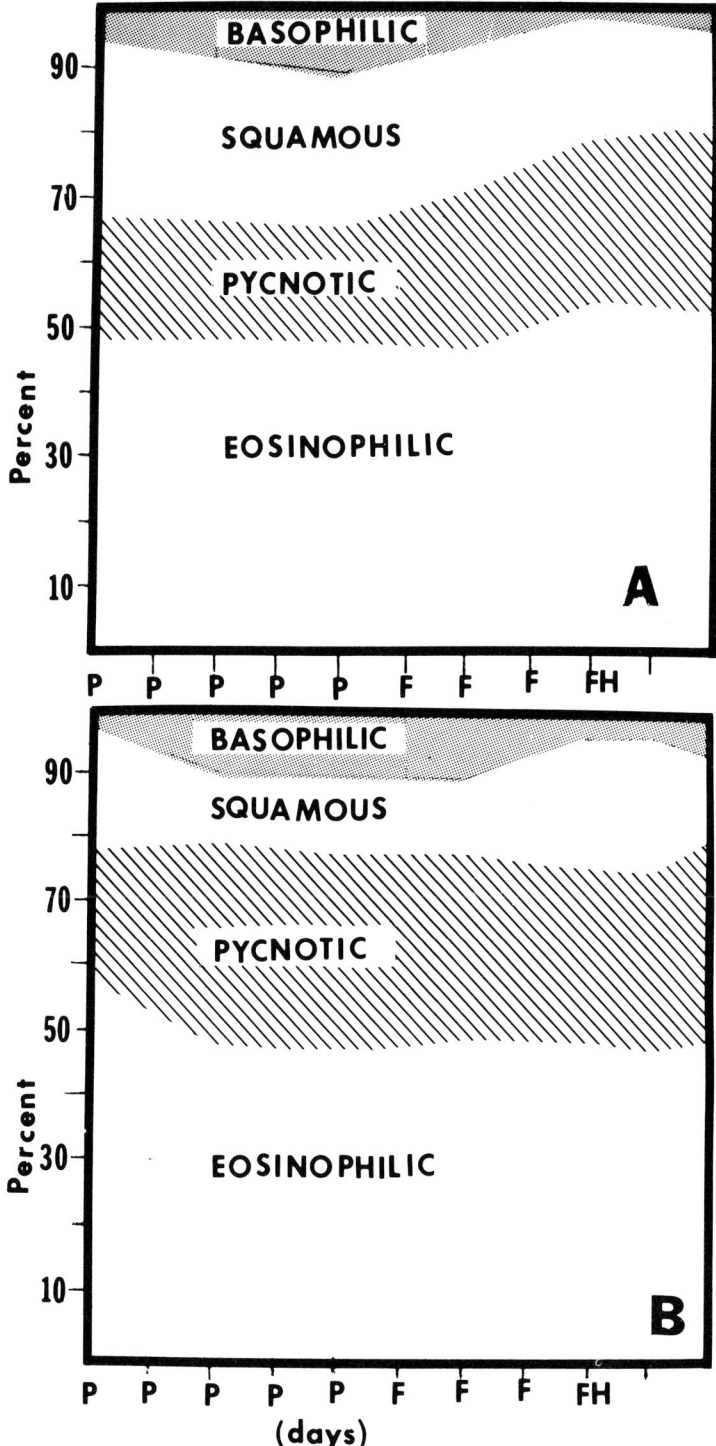

Figure 1. Cytograms of exfoliate vaginal cells during ovulatory treatment during (a) the spring and (b) the summer test periods. (P = progesterone; F = FSH; FH = FSH followed by HCG).

TABLE II

In vitro Fertilization of Squirrel Monkey Ova Recovered from the Follicle 12 h After HCG

Factor	
Total cultures	41
Total ova	58
Ova maturing/total ova	26/46
Percent maturing	56.5
Ova fertilized/ova maturing	17/23
Percent fertilized	74.0

seasons of 1975 thrice daily observations were made for one hour, 0800, 1300 and 1600, to study the behaviour of the animals in the outdoor cages.

RESULTS AND DISCUSSION

The effectiveness of the various ovulation regimes is shown in Table I. The distinct seasonality can be seen in the differing responses observed between the spring and summer seasons. During the summer season there was a 79.5% decrease in the ovulatory response compared with the high months of March and April. The administration of varying levels of HCG, ranging from 500 to 1,500 iu, had no effect on the ovulatory response, whereas the amount of FSH did influence summer response. From these data we postulated that the basic ovulation regime was minimal for ovulation and that during the summer season this regime become subminimal with respect to the FSH. Confirmation of this hypothesis is seen in Figures 1 and 2. Figure 1 indicates the proportion of various cell types in the vaginal smear in the spring (a) and summer (b) after treatment with an FSH-HCG regime. With the administration of FSH, and subsequent follicular growth, there is an increase in the percentage of eosinophilic and pyknotic cells during the spring. However, during the summer season this level of FSH fails to alter the percentages of these cells, indicating a subminimal dose. Figure 2 illustrates the effect of the FSH-HCG regime on the number of attempts to mount, penile displays and erections of associating males during two spring treatment sessions (A and B) and two late summer treatment sessions combined (C). Again the stimulatory effect of the FSH-HCG regime is seen in the spring but is nearly absent during the summer months.

The fertility of the ova resulting from induced ovulation can be examined both *in vitro* and *in vivo*. Table II indicates the fertilizability of ova recovered from the follicles 12 h after HCG injection and subjected to *in vitro* fertilization. It will be observed that of mature oocytes, 74% of the ova were fertilized. *In vitro* fertilized ova have been developed to the eight cell stage in culture and to morula stage after *in vivo* culture in rabbit oviducts. During the first year, with *in vivo* fertilization, twenty-four females were housed with four males and half the animals subjected to ovulation induction. Eight pregnancies were obtained of which three went to term. Pregnancies occurred in both treatment and control groups.

Incomplete results from the 1976 summer season, with 25 females, indicate eight probable pregnancies occurring in both treatment and control groups. In this year treated females received five days of FSH (1 mg) and 500 iu HCG.

These results from both *in vitro* and *in vivo* fertilization verify the

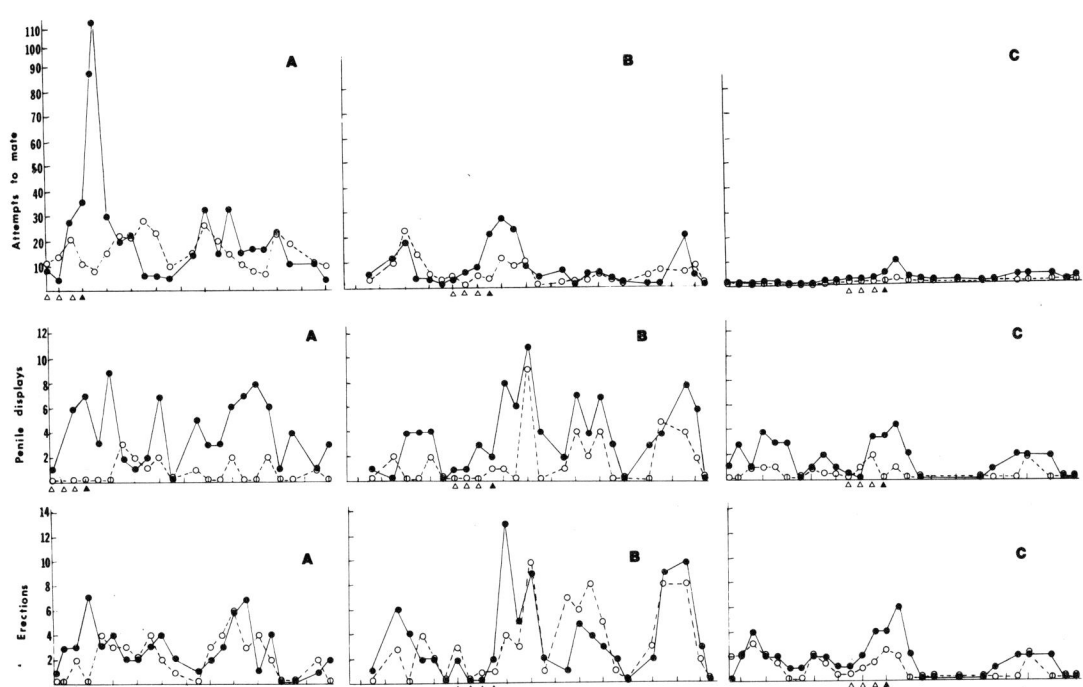

Figure 2. *Behavioural responses during the treatment periods (A, B = combined data from spring sessions; C = combined data from last two summer sessions; O = control animals; ● = treated animals; △ = FSH injections of the ovulatory treatment; ▲ = HCG injections of the ovulatory treatment).*

fertility of the induced ova and further demonstrate the feasibility of inducing ovulation and obtaining pregnancies in the anovulatory season.

The problem of a high rate of embryonic and foetal loss, as well as a high neonatal death loss reported by others, remains to be solved. These factors are presumably stress-related, and additional research is needed on the housing and management effects on such survival.

The usefulness of laparoscopy in such studies for ovulation detection and pregnancy diagnosis is confirmed and the lack of a deleterious effect on pregnancy confirms our earlier findings.

It would appear that captive breeding on *Saimiri sciureus* is practical and feasible despite the seasonal breeding that these animals experience. Using gonadotropin therapy during the anovulatory season can stimulate follicular growth and ovulation, which can result in subsequent pregnancy. Problems of embryonic and foetal death as well as neonatal survival have not been examined in the present studies but require additional work before a complete programme of production can become reality.

SUMMARY

Studies over the past seven years indicate a seasonal response to external gonadotropins in *Saimiri sciureus*. This effect can be overcome by increasing the levels of FSH during the anovulatory season yielding ova which can be fertilized both *in vitro* and *in vivo*. The laparoscope can easily be used as a tool for diagnosis of ovulation and pregnancy with no adverse effects on the pregnancy. The usefulness of *S. sciureus* as a research primate can be greatly extended by a sound conscientious programme of captive breeding coupled with research on the nuances of reproduction and newborn survival of these species.

ACKNOWLEDGEMENTS

The author wishes to acknowledge the expert collaboration of Drs. R.M. Harrison, T.J. Kuehl and S.J. Jarosz with the reported projects. This work was supported by NIH grant number HC07534 and a grant from the National Foundation-March of Dimes. Publication approved by the Michigan State University Agricultural Experiment Station Number 7813.

REFERENCES

Bantin, C.C. (1966). *J. Inst. Anim. Tech.*, 17, 66-73.
Bennett, J.P. (1967a). *J. Reprod. Fertil.*, 13, 357-459.
Bennett, J.P. (1967b). *J. Endocrin.*, 13, 473-474.
Clarkson, T.B., Bullock, B.C., Lehner, N.D.M. and Feldner, M.A. (1968). *In* "Animal models for biomedical research", pp. 64-87. National Academy of Sciences, National Research Council, Washington, D.C.
Dukelow, W.R. (1970). *J. Reprod. Fertil.*, 22, 303-309.
Dukelow, W.R. (1975). *J. Reprod. Fertil. Suppl.*, 22, 23-51.
Dukelow, W.R., Jarosz, S.J., Jewett, D.A. and Harrison, R.M. (1971). *Lab. Anim. Sci.*, 21, 594-597.
Dukelow, W.R. and Kuehl, T.J. (1975). *In* "La Fécondation", (C. Thibault, ed.), pp. 67-80. Masson Publishers, Paris.
DuMond, F.V. (1968). *In* "The Squirrel Monkey", (L.A. Rosemblum and R.W. Cooper, eds), pp. 87-145, Academic Press, New York.
Harrison, R.M. and Dukelow, W.R. (1973). *J. Med. Prim.*, 2, 277-283.
Hupp, H.W. (1972). *In* "Medical Primatology I", pp. 100-104, Karger, Basel.
Jarosz, S.J., Kuehl, T.J. and Dukelow, W.R. (1977). *Biol. Reproduction*, 16, 97-103.

BREEDING *ALOUATTA CARAYA* IN CENTRO ARGENTINO DE PRIMATES

O. COLILLAS AND J. COPPO

**Centro Argentino de Primates,
Sefrano 661, Buenos Aires 1414, Argentina.*

INTRODUCTION

In 1973, CONICET, The National Research Council on Technical and Scientific Research in Argentina, decided to start a centre for the breeding and rearing of primates.

The place chosen was in the Province of Corrientes, near the forests where there were known to be monkeys.

During the time the installations were being prepared, some field studies and experiments were carried out on the type of habitat, food, and sexual cycles of the *Alouatta caraya*. It was decided to start with this type of monkey as they were easily obtainable in districts close to the Centre.

Although a great deal has been written about the howler monkey (e.g. Altmann, 1959; Cabrera and Yepes, 1960; Carpenter, 1934; Collias and Southwick, 1952; Crespo, 1954; Pope, 1966; Southwick, 1962) our aim was to update this information from our own experience. It would also be a way for the personnel we had engaged to work at CAPRIM to get valuable training.

HABITAT

We first had to establish the actual limits where the *Alouatta caraya* lives in the north (middle and eastern sectors) of our country.

We travelled through the provinces of Misiones, Corrientes, Chaco and Formosa along the principal roads, turning down side roads running perpendicular to these, every 20-30 km, so as to cover the greatest area possible in each province (Figure 1). Every 20-30 km we would question the people living in the district and whenever possible get them to come with us so as to verify the existence and variety of the monkeys which inhabited that particular area. This method of travelling through the provinces was decided on due to the many fairly good roads which exist close to the forest and not far from the water courses.

So we were able to prove that monkeys inhabit an area demarcated to the south by parallel $29°$ in the province of Corrientes, $27°$ in the Chaco and to the west by meridian $60°$.

The *Alouatta caraya* is to be found in all this area, is the only species in the province of Corrientes, co-exists with *Cebus* sp. (Cai Misionero) in Misiones and with *Aotus* sp. (Miriquinha) in Formosa and Chaco. A rough demographic density was made (Figure 2).

*Consejo Nacional de Investigaciones Científicas y Técnicas (CONICET).

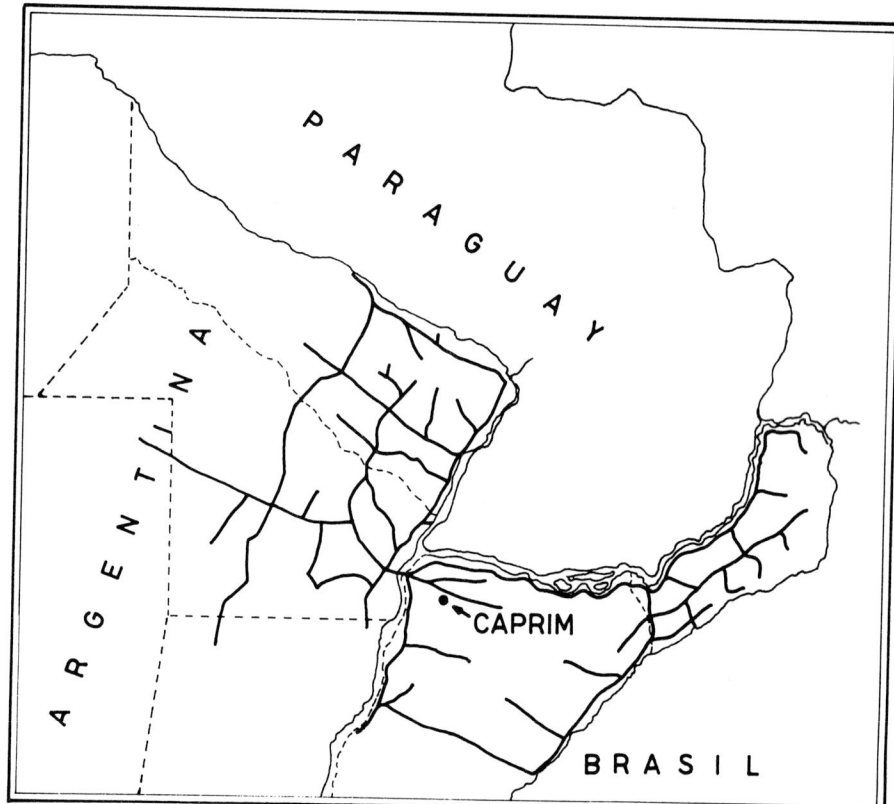

Figure 1. Roads followed to delimit the habitat of monkeys in northeast Argentina.

The troops of *Alouatta* were found living in the inland forest close to streams in the tree-tops of all tall trees. During the summer season, when it does not rain so often, we found the monkeys more frequently close to the banks of the larger rivers where the trees were smaller.

It was observed that each troop inhabited an area of about one hectare, but we did not notice any connection between the size of the area where they lived and the size of the troop. Their gregarious behaviour, due to need to get food, does not stop them from staying for a long period in the same district if this provides a good dwelling place. Therefore it is usual to find howler monkeys living for years in areas close to villages where they are given food by the local inhabitants.

Their way of travelling in the forest is by balancing and swinging from the branches of trees. It is rare for them to jump and they only do this in case of emergency, jumping downwards towards another branch or to the ground. The distance covered in these cases is never more than 4 m. These observations were carried out to design the building of the corrals.

SIZE OF THE TROOPS

We have observed 40 troops throughout 18 months. The number of animals which make up a troop varies with the cold or hot season (Table I).

They are more numerous in winter, when we saw up to as many as 20 in one group ($\bar{x}$: 12.07 ± 3.53).

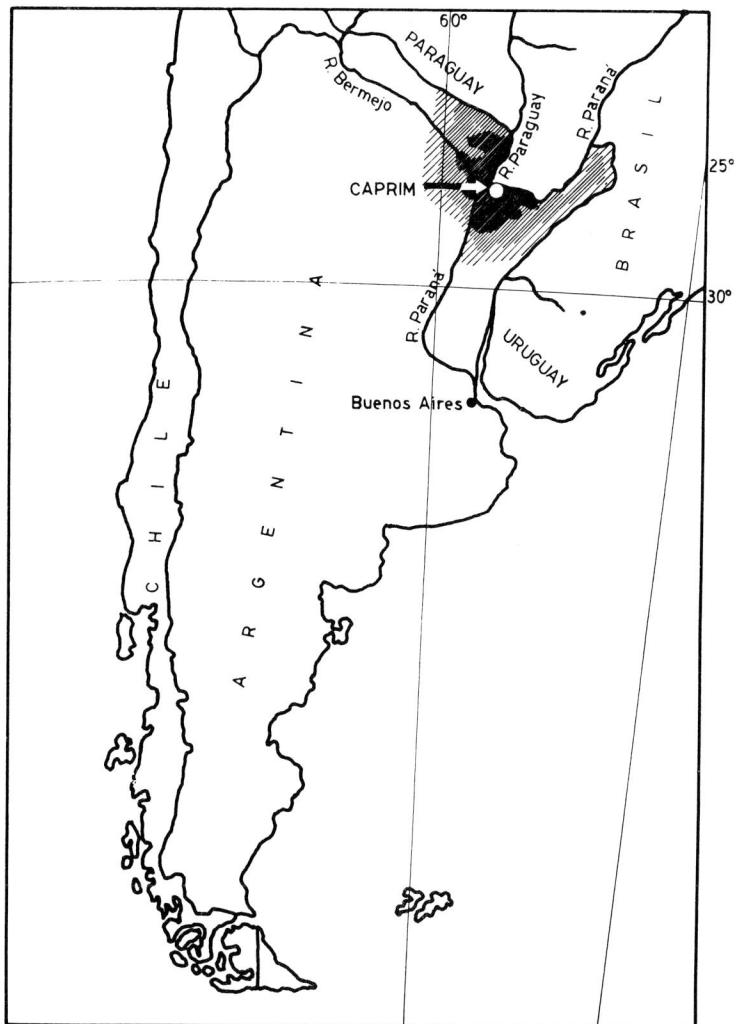

Figure 2. Habitat of Alouatta caraya *in northeast Argentina. Rough demographic density.*

In the summer the groups separate and usually consist of fewer animals ($\bar{x}$: 5.5 ± 0.5). We also frequently found a pair on their own, living in quite an extensive area of the forest.

In every troop there is a male leader who tolerates other males, generally younger than he is, and they are allowed to mate with the females. Any strange male is driven off by the leader.

BREEDING SEASON

Every time we went into the forest, whether to hunt or simply to observe, we observed if there were any off-spring or if the animals were mating. For this reason any females captured were examined to see if they were pregnant and urine was extracted for an immunological pregnancy test.

In the 18 months our study lasted there were two hot and one cold season.

TABLE I
Sexual Behaviour in Different Kinds of Enclosures

Group		Mating	Pregnancy	Time Lasted	Season
Outdoor Cages	A 1M/3FM	1	-	3 months	Summer
	B 2M/3FM	-	-	3 months	Winter
	C 1M/2FM	-	-	3 months	Spring
Small Corrals	D 2M/2FM	-	-	4 months	Autumn/Winter
	E 2M/6FM	-	-	4 months	Autumn/Winter
Large Enclosure	F 1M/6FM	-	2	4 months	Summer
	G 1M/6FM	-	-	2 months	Autumn
Indoor Cages	3 Pairs	-	-	4 months	Summer/Autumn
	3 Pairs	-	-	2 months	Autumn/Winter

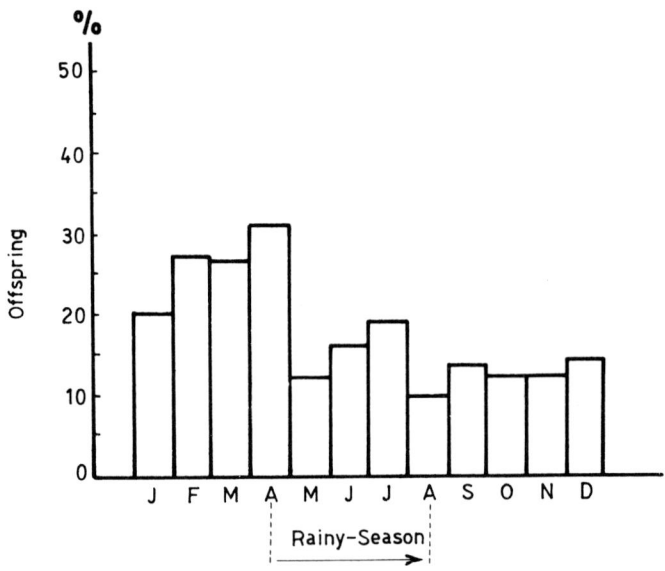

Figure 3. Monthly percentage of offspring observed during 1975.

Throughout this time we observed off-spring with their mothers and, among the animals captured, found females who were pregnant, but this was seen with the greatest frequency at the end of the summer (February, March and April) and the beginning of autumn just before the start of the rainy season (Figure 3).

We were never able to see any mating in the forest, and the existence of isolated couples was more frequently observed during the hot season but, as already stated, this is also the time when the troop breaks up.

CAPTURE OF ANIMALS

Various methods were tried to capture the animals. The people working at CAPRIM were in charge of this job and took into account the former observations to choose the places and methods of capture.

We tried out every means of trapping the animals told to us by the local

TABLE II
Sexual Cycle of Alouatta caraya *Studied by means of Vaginal Cytology*

No.	D.C.	S.C.P. (days) From capt.	S.C.P. (days) Between	%	R.B.C. (days) Before scp	R.B.C. (days) After scp	Length	P.V.E.
1	Jan	8		60				
			18	60				
			18	58				
				-				22 d
				68	12		1	
			13	57		5	1	
2	Jan	23		65				
			22	75				
			24	70				
3	Jan	11		65		7	1	
			19	65				
			17	71	10			
				-				
				65				
4	Mar	17		48	7	7	1	
			22	60	15		2	
			19	54		4	1	
5	Mar	10		53	8		4	
			22	70				
			23	69		10	5	
6	Mar	17		55		8	1	
				-				14 d
				69	7	7		
	x	14.2	19.72	62.8	9.83	6.85	1.9	
	S		± 0.99	± 1.61	± 0.97	± 0.75		

D.C. = date of capture; S.C.P. = squamous' cells peak; P.V.E. = persistent vaginal estrus.

inhabitants but without success.

The Cap-Chur system, using 1 ml syringes with nicotine alkaloids, xylazine or ketamine hydrochloride was useless. The same amount of these sedatives given intramuscularly produced a deep sedation in the animals. We believe that part of the content in the dart-like syringes is lost in firing as we tried the injector mechanism with a coloured fluid and it worked perfectly.

We also tried the usual kind of trap-cages used to capture animals that inhabit the forest but the *Alouatta caraya* would not enter them.

After these fruitless efforts we decided we would have to use more primitive methods. One of these was to surround an isolated tree where there were one or more animals, shaking it vigorously if the trunk was fairly slender, or setting traps among the branches if it was massive so that the animals would fall and be captured.

Once the monkeys had been caught they were given a sedative (benzodiazepine 3 mg/kg) and placed in transport cages.

During the hot season the animals are to be found in fairly open forest and are well nourished. During the winter the animals go to denser forests with taller trees. We have even discovered them living in the middle of reedbeds where they are difficult to catch. At this time of the year the temperature is very low, rain is frequent and the monkeys are frequently found to be suffering from severe anaemia, so we decided it was better to try and capture them in the summer.

During 18 months, 110 *Alouatta caraya* (Table II) have been captured on the banks of the rivers Paraná and Paraguay and also in places further inland close to CAPRIM. All the animals were kept in quarantine and submitted to the usual clinical and haematological examinations. During this period they were given tuberculin PPD tests and vaccinated against TBC, measles and rabies.

In order to estimate the age of captured howlers, dental-age methods of relative aging are combined with the body weight data.

Forty of the original number of *Alouatta caraya* captured are still alive. We found a 30% mortality rate due to inexperience in handling these animals and the inadequate protection given them during the exceptionally severe winter of 1975. Other animals escaped, died from conditions other than pneumonia, or were used in different studies.

TYPES OF CAPTIVITY

We tried three different kinds:
1. Simulated free-range: small and large enclosures.
2. Outdoor colony cages.
3. Indoor individual and paired cages.

Simulated free-range

To test their efficacy we had small corrals built, 300 to 400 m^2 in size, surrounded by corrugated iron 2.5 m high. These enclosures surrounded small clumps of trees, leaving a clear area of about 4 m around them. We decided not to use any kind of door which would have been a way of escape but used a system of movable ladders instead.

Our idea at first had been to construct the larger enclosures using a system of corrugated iron and electrified wiring. After several trials we built a fence 1.7 m high, surrounding an enclosure of 1 hectare. For three months the animals were kept there without an escape. Thinking this structure was fool-proof we increased the number of animals enclosed in this corral. The result was a fight between two leaders and the subsequent escape of the whole troop belonging to the loser. As a result of this the large corrals were re-modelled and the fence raised to a height of 2.5 m. We eliminated the electrified wiring.

Outdoor colony cages

Four cages of 60 m^3 and another of 27 m^3 were built and placed in such a way that they got direct sunlight all day, with a corrugated iron wall facing south as a protection from the cold wind.

Indoor individual and paired cages

When a special precinct was ready, cages of 1 m^3 were built of wood and wire with a movable backwall. Inside this precinct there is an even temperature of 25° and artificial lighting is present for 14 hours from 0700 to 2100 h. This is produced from a fluorescent ceiling light giving 10 ft candles of illumination. Every three hours the place is ventilated.

FEEDING

When the animals are first enclosed they refuse any sort of food in pellet form. To avoid suffering it was decided to try different kinds of food. Ultimately we decided on a mash of sweet potatoes, pumpkin, eggs, powdered milk, honey, wheat or soybean flour, and carrots, with added vitamins (VIONATE). The total diet is 1.13 Cal/g, 10.41 g % protein, 2.12 g % lipids, 13.06 g % carbohydrates, 3.66 g % minerals and 69.35 % water.

Each animal ate 400 g of this diet per day during summer and an extra 150 g per day during the winter months. In the afternoon, to give them a more varied diet, any fruit that is in season (apples, grapes, mangoes, bananas) as well as green leaves or a pudding made of maize or wheat, is added on alternate days. The sites for the corrals were chosen to provide the animals with a variety of vegetation which they might eat.

SEXUAL BEHAVIOUR IN CAPTIVITY

During the time in isolation the animals were carefully observed so as to notice any tendency to form groups or to pair off.

Outdoor cages

Three groups were formed (A, B, C) consisting of 1 male/3 females, 2 male/3 females, and 1 male/2 females respectively. Each group was observed to see if they mated, and their vaginal contents were examined. After 3 months the females were captured and an immunological pregnancy test was performed.

On only one occasion did we observe a mating, but the female did not become pregnant. No traces of spermatozoa were found in the vaginal smears.

During all this time no female became pregnant.

Small corrals

Two troops (D and E) were placed in 2 small enclosures of 400 and 800 m^2.

Group D: 2 male/2

After 4 months without being able to observe any mating, with that winter being extremely long and cold (leaving the trees with little foliage and so giving poor shelter), it was decided to remove the animals. No females were found to be pregnant.

Indoor cages

Pairs of animals which had just finished their quarantine were placed in indoor cages. All the females' vaginal contents were examined periodically in order to study their ovarian cycle in this type of captivity and to see if there were any spermatozoa.

During this time, 4 months in all, we did not observe mating and none were pregnant. We feel, however, that the animals have not had sufficient time to get accustomed to their new surroundings and to their treatment.

Large enclosures

When the first of the large enclosures was completed, in September 1975, a small troop was placed there. Some minor defects in the fencing which had been overlooked during the bi-weekly inspection enabled 16 monkeys to escape. Most of them were females who appeared to be cleverest at getting away. Only in January of this year was it possible to get a group in this enclosure. This was made up by troop F, 7 adult animals (1 male/6 female) and 3 juveniles.

It was immediately noticed that 3 of the adult females left the main group to huddle together 80 m away from the others. We were therefore able to form troop G (1 male/3 female). Male leader F would not tolerate the presence of male G, fighting him until he was finally so seriously injured that he died. When another male was put into the enclosure to replace male G, the new leader managed to escape with the whole of troop G, although we had thought the fencing impassable. For this reason the large enclosure was remodelled and the height of the fence raised to 2.5 m. We now have 2 large enclosures of 5,000 m^2 and 3,000 m^2 each.

In the first is group F with 2 other new females and 7 juveniles. In the second we have recently put 12 animals (a new group G) consisting of 1 male/ 6 females and 7 juveniles of both sexes. None of these animals has been recaptured since they were placed there.

During the last summer and autumn the females howled loudly and frequently in a way we had never heard before.

The result of this behaviour was the appearance of numerous *Alouatta* in the forest close to CAPRIM where monkeys had not been seen for several years. When captured these males proved far fiercer and more sexually aggressive than any we had captured before.

After troop F had been living in captivity for 3 months it was noticed that 2 females seemed to be pregnant, based on the increase in the size of their bellies and breasts.

As a result of 18 months of work we have come to the following conclusions:
1. Outdoor colony cages do not appear to be useful for breeding *Alouatta caraya*.
2. Small corrals need better shelter than the trees to protect the monkeys during the cold season.
3. A male leader *caraya* will not tolerate another in the same enclosure if he has characteristics similar to his own, although they are living at some

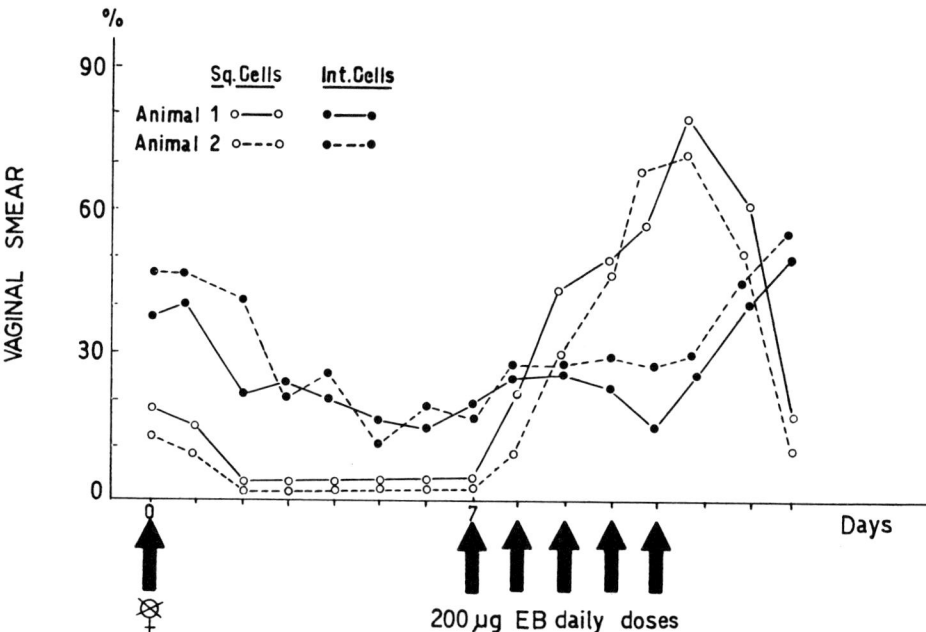

Figure 4. Vaginal response to estrogenic stimulus. (Sq: squamous cells; Int: intermediate cells; EB: estradiol benzoate: gonadectomy)

distance from each other and each has his own group of females.

4. In the large enclosures the behaviour of the animals would appear to be no different from that when living free in the forest except that they get accustomed to the presence of man.

5. The females seem to emit a special sort of howl which attracts the males.

6. Electrified wiring at low voltage proved to be no obstacle when the animals are desperate to get away.

REPRODUCTIVE CYCLES

We have found few references on the reproductive cycles of the *Alouatta caraya*. Dempsey described a cyclical activity of ovaries based on a study of gonads of dead howler monkeys. Our own experience, after having observed offspring and pregnant females throughout a year, leads us to believe that female *caraya*, just as the *Alouatta palliata* (Carpenter, 1934) is a polyoestrous monkey. With the aim of corroborating this supposition it was decided to study the reproductive cycle of the female *Alouatta caraya* and we divided the work into two parts:

1. To do a year's study on the microscopical morphology of the ovaries, always trying to discover recent corpus lutea and developing follicles.

We studied the gonads of dead animals from December 1974 to February 1976. During this time 22 ovaries from 11 animals were processed. In 9 of them recent corpus lutea were found and the other 12 were in different stages of involution. Developing follicles, and sometimes tertiary follicles, were seen in all the ovaries.

2. To do a study on the vaginal contents of animals which had entered

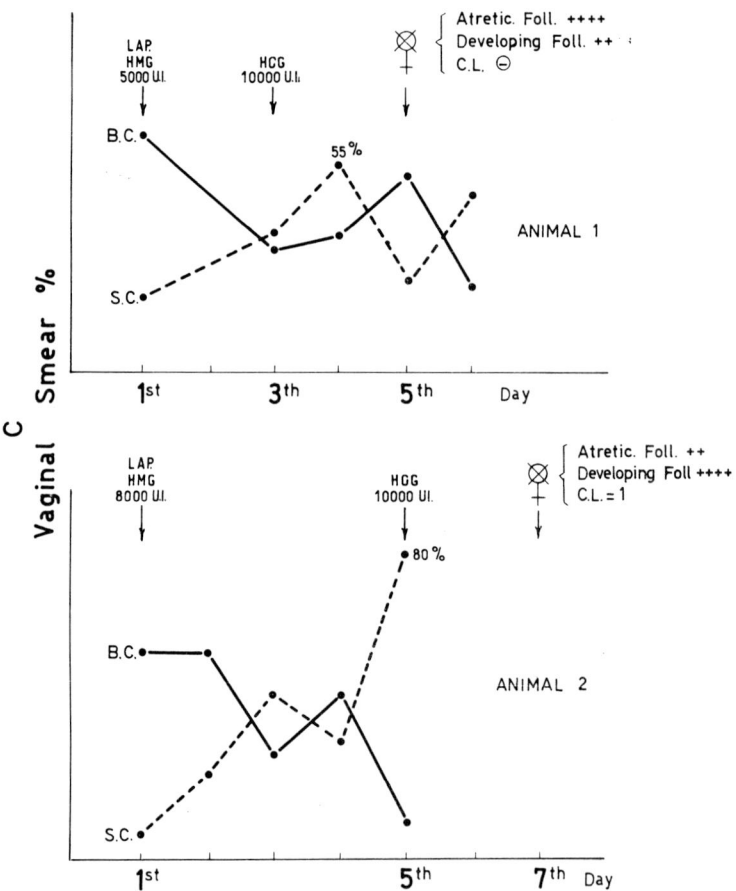

Figure 5. *Vaginal response to HMG and HCG stimulus. (Lap: laparotomy; CL: corpus luteum; gonadectomy)*

CAPRIM. Samples were taken every two days and dyed using Shorr's method (Shorr, 1941). Strict control was kept to discover any increase in the number of squamous cells.

No samples taken showed any rise of squamous cells nor was cyclicity seen in the other vaginal cells.

As we were unable to obtain perfectly stained specimens we decided to interrupt the study and consider three possibilities:
1. Shorr's is not a useful technique for this species.
2. Animals held in captivity remain in anoestrus.
3. Squamous cells are not a good index for ovulation detection in female *caraya*.

It was therefore decided to do other short studies to test this last hypothesis using a modified staining of Papanicolau.

Two adult females were spayed and after 7 days they were given a daily dose of 200 µg EB for 5 days. After the third day of being given the oestrogen there was an increase in the number of squamous cells over intermediate cells. This increase was very noticeable in subsequent days, reaching 75% after the last injection (Figure 4). Intermediate cells and WBC persisted in all samples.

It was decided to test the effect on the vaginal epithelium when an increase of ovarian activity is obtained with HMG and HCG stimulation. Two females were chosen and daily vaginal smears were taken to check for an increase of squamous cells.

On Day 1 the monkeys had a laparotomy to verify that there was no recent ovulation. That same day they were given 5,000 iu and 8,000 iu of Pergonal (Searle Laboratories) respectively to monkeys 1 and 2.

Female 1 was given 10,000 iu of HCG on Day 3 and female 2 the same dose when the percentage of squamous cells reached about 80 in the vaginal smears.

Both females were submitted to a second laparotomy, 48 hours after having been given HCG, and ovaries were obtained for microscopical study (Figure 5).

Results

Monkey 1 - ovaries. A great number of atretic follicles were found, only a few were in development; there was no recent corpora lutea. The interstitial tissue showed signs of intense stimulation.

Monkey 1 - vaginal smears. A moderate rise in the percentage of squamous cells (55%) was obtained, with a persistence of intermediate cells.

Monkey 2 - ovaries. Some atretic follicles were found, many others were developing and there was also a recent corpus luteum. Just as in animal 1, interstitial tissue showed signs of intense stimulation.

Monkey 2 - vaginal smears. Five days after being given HMG a net increase in squamous cells (80%) was obtained but there were still a number of intermediate cells and leucocytes.

Comments

The staining, when done according to Pap's modified technique, was easy to handle and samples easily interpreted.

It would seem that squamous cells could be a good index to detect oestrogen increase in female *caraya*. The persistence of intermediate cells and even leucocytes during the peak of squamous cells is in accord with Dempsey.

Therefore, the results of these two short studies allowed us to begin research on the reproductive cycle of the female *Alouatta caraya* in captivity.

The plan is of a year's duration with the entry of 3 new animals every 3 months until there are at least 12 monkeys. Single adult females are kept in cages of 1 m^3 together with a single adult male. Samples of the vaginal contents are taken every two days by means of glass pipettes. Before the sample is taken the external genitalia are washed and the pipettes sterilized. The samples are stained according to the modified Pap technique and four kinds of cells are counted: (i) squamous cells; (ii) basophilic superficial cells; (iii) intermediate cells and (iv) parabasal cells. We counted 200 elements in each sample and WBC, RBC and spermatozoa were also registered.

Some of the females were submitted to a laparotomy during or after the squamous cells' peak to detect follicles near ovulation.

External genitalia and breast changes were observed so as to compare with cytological changes.

The animals were sedated with ketamine hydrochloride 5 mg/kg.

Results

During the last five months (late summer, and early autumn) it was noted

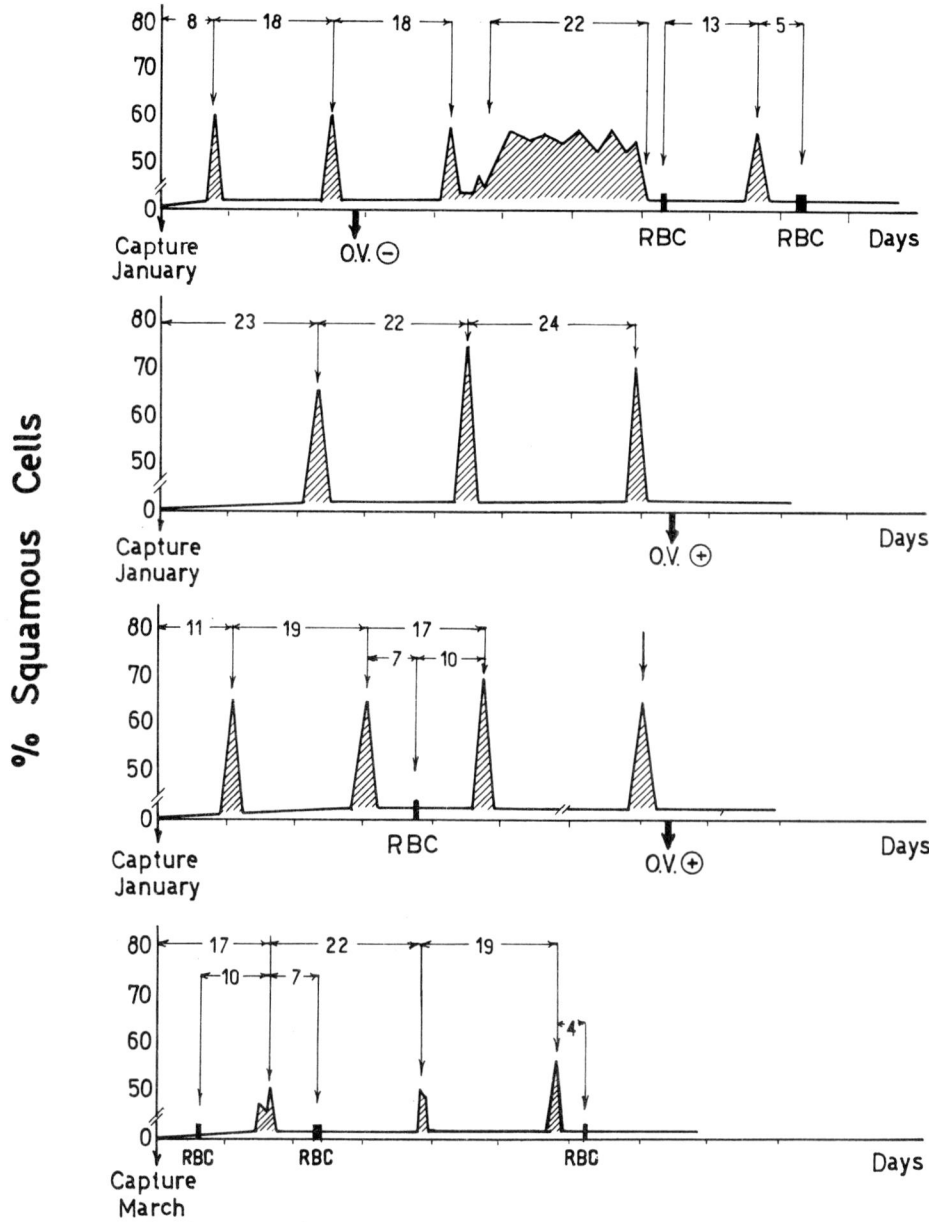

Figure 6. *Cytological vaginal cycle of* Alouatta caraya. *(OV: ovulation detection by laparotomy)*

that (Figure 6 and Figure 7):

1. When the animals were first examined after capture, smears always showed parabasal cells and WBC.

2. The first smear to have a predominance of squamous cells was seen 8 to 23 days after capture.

3. After the first peak of squamous cells this prevalence was repeated every 17 to 24 days ($\bar{x} = 19.72 \pm 0.99$) but without the complete disappearance of the intermediate cells or WBC. On these occasions the percentage of squamous

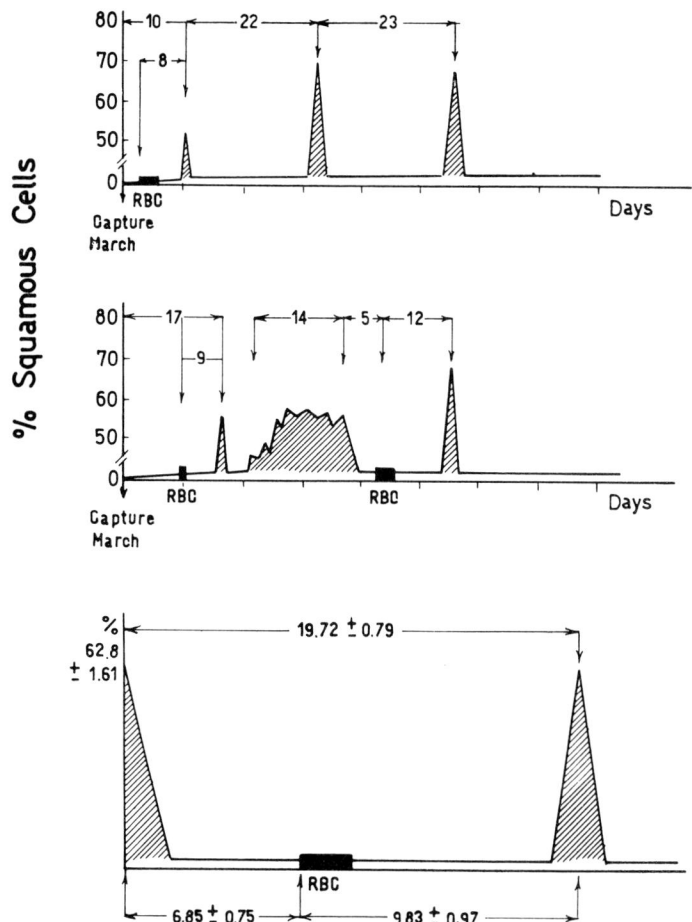

Figure 7. *Cytological vaginal cycle of* Alouatta caraya. *(OV: ovulation detection by laparotomy)*

cells range from 48% to 75% ($\bar{x}$ = 62.8 ± 1.61) and lasted between 48 and 72 hours.

4. Sometimes many erythrocytes were found in the slides. These appeared 4 to 10 days ($\bar{x}$ = 6.85 ± 0.75) after the squamous cell peak, that is 8 to 12 ($\bar{x}$ = 9.83 ± 0.97) days before. On one occasion, microscopical bleeding was discovered 48 to 72 hours after an adult female animal had been spayed.

5. On two occasions two animals were found to have a predominance of squamous cells (58%) for 14 and 22 days and, after a microscopical bleeding, returned to their normal cycle.

6. On the days previous to or immediately after the squamous cells peak the intermediate and parabasal cells were predominant. At the end of the cycle small cells similar to the basal or endometrial cells appeared.

7. No changes were noticed in the genitals or breasts which could be correlated with the cytological changes.

COMMENTS

It would seem that the *Alouatta caraya* continues to be a polyoestrous animal even in captivity and that its cycles can be assessed by vaginal cytology.

The stress of being captured apparently interrupts this cycle, which returns after a variable period or a microscopical genital bleeding. This microscopical bleeding is due (as in women) to the fall of the ovarian steroid levels. It lasts for such a short period that it can easily pass unnoticed. The reproductive cycle of the female *Alouatta caraya* in captivity lasts roughly 20 days and ovulation should take place immediately after the peak of the squamous cells. The follicular period appears to last 9 days and the luteal 7 days. The end of one cycle and the start of the next is seen by the microscopical bleeding already mentioned. When, due to its brief duration, it has not been detected, the presence of endometrial and basal vaginal cells could be the sign of the end of the cycle.

REFERENCES

Altmann, S.A. (1959). *J. Mammal.*, 40, 469.
Cabrera, A. and Yepes, J. (1960). Mamíferos sudamericanos. Historia natural Ediar, Tomo I, Campanía Arg. de Editores, Bs.As. 106-109.
Carpenter, C.R. (1934). *Comp. Psychol. Monog.*, 10, 1-168.
Collias, N. and Southwick, C.A. (1952). *Proc. Amer. phil. Soc.*, 96, 143.
Crespo, J.A. (1954). *J. Mammal.*, 35, 117.
Malinow, M.R., Stahl, W.R., Maruffo, C.A., Pope, B. and Depaoli, R. (1966). *Primates*, 7, 433-447.
Pope, Betty L. (1966). *Am. J. Phys. Anthrop.*, 24, 361-370.
Schultz, A.H. (1960). *Proc. zool. Soc. London*, 133, 23-390.
Shorr, E. (1941). *Science*, 94, 545-546.
Southwick, C.H. (1962). *Ann. N.Y. Acad. Sci.*, 102, 436-454.

ON THE KEEPING, FEEDING AND BREEDING OF LEAF MONKEYS IN THE THÜRINGER ZOOPARK OF ERFURT

D. ALTMANN

Thüringer Zoopark, Erfurt, East Germany.

Most of the species of leaf monkeys are threatened with extinction in their African and Asian habitats. Keeping them in zoological gardens is complicated, because these monkeys, which have a multi-chamber stomach like the ruminants, preferably eat blossoms, buds and leaves. They need quite a lot of them. Dry leaves can be used as substitute only to a limited extent. The sprouts of cherry tree branches and other fruit trees from around December on during the winter season, proved to be good for the monkeys. It is particularly advantageous to deep freeze the leaves packed tightly in plastic sheet bags. Oak trees, willow bushes, rose trees, raspberry and blackberry bushes, beech, lime, birch, ash and fruit tree woods turn out to be appropriate but, naturally, without previous treatment with contact insecticides.

Various sorts of bindweed remain in leaf until far into the autumn. As a rule, salad, lettuce and, in particular, endive can be kept fresh until December, provided that they are kept in adequate cold frames or cold protected areas. They also serve as leaf nutrition. Sprouted grain at a length of 6 to 8 cm is used likewise, but only to a limited extent.

Other staple foods for leaf monkeys are carrots and apples. Bananas are given only to a limited extent and in a green or half-green state, because of their high sugar content.

Besides other fruits, tropical and subtropical fruits, and some sorts of vegetables, a certain amount of grain feed is given, consisting of wheat, maize (kibbled), oats, peanuts and kernels of other kinds of nuts, sunflower seed, linseed and coarse soybean meal.

In order to satisfy the high protein requirement for the animals and also to supply carbohydrates, fats, vitamins, minerals and nutrients, the leaf monkeys living in Erfurt are being given daily concentrate dumplings consisting of the following substances:

Oatmeal (if not available, oat flakes)	35%
Dried milk	25%
Eggs, raw	25%
Butter	25%
Rachitin (mineral substance)	2.5%
Afarom (nutrient mixture)	2.5%

In addition, the following vitamin compounds are added:

Ursoselevit AD$_3$EC	(VEB Jenapharm - Serumwerk Bernburg)
Vitamin-B-Complex	(VEB Jenapharm - Serumwerk Bernburg)

Ascorvit	(VEB Jenapharm)
Ursoselevit (Vitamin - Selenpräparat)	(VEB Jenapharm - Serumwerk Bernburg)
Rachitin	(VEB Chemisch-Pharmazeutisches Werk Oschersleben/Bode)
Afarom	(VEB Jenapharm - Serumwerk Bernburg)

The leaf monkeys, in many respects, show a behaviour that is different from other species of monkeys. This especially applies to the order of dominance. Whereas the quarrels about dominance are accompanied by biting in the case of macaques, guenons, mangabeys and baboons, the mental pressure plays the leading role in the case of the leaf monkeys. Such a feeling of oppression with the low-ranking males can be so pronounced that they hardly dare sit on the perches in the cages.

High-ranking females can become lower in rank due to the failure to move when pregnant. They are not allowed on the perches and, to a certain degree, are kept away from the feeders by the females that are higher in rank. This can be changed by a skilled keeper who is in the cage where he mimics the turning motions of the species of monkeys concerned. Immediately, the female monkeys that have become lower in rank will, under the care of the keeper who is being considered a high-ranking fellow monkey, feel at ease and will take food, which otherwise might not be the case.

The breeding of rare species of leaf monkeys under the care of man is of vital importance for the clarification of the relationships between individual species of monkeys. Hence, the adult dusky leaf monkeys, predominantly grey coloured and showing non-pigmented, spectacle-like circles around the eyes, are distinctly different from John's langurs, which are predominantly coloured black-brown and do not show any spectacle formation around the eyes. Young dusky leaf monkeys are of a golden brown colour, but they show the non-pigmented appearance around the eyes, looking like spectacles. The young John's langurs are likewise of a golden brown colour and show such a spectacle formation around the eyes. However, it disappears totally at the time when the coat changes its colour.

These observations were made with the John's langurs born in 1972 in the zoo of Erfurt. The birth in question was the first one in Europe.

The Thüringer Zoopark, Erfurt possesses an interesting collection of leaf monkeys: East African (one of the largest breeding groups in Europe) and West African black-and-white colobuses, Hanuman's langurs, John's langurs (outside the Indian continent where they are strongly threatened, only in San Diego, California, USA and Erfurt, GDR), Douc langurs (only a few groups in Europe and North America) and proboscis monkeys (only a few groups in Europe and North America).

BREEDING BABOONS IN UGANDA AND CAMBRIDGE

B.A. BAKER, G.F. MORRIS AND T.D. COWAN

*Medical Research Council, Dunn Nutritional Laboratory,
Milton Road, Cambridge, UK.*

INTRODUCTION

In 1969 the Medical Research Council Child Nutrition Unit in Kampala, Uganda, under the direction of Dr. R.G. Whitehead, was investigating the causes and treatment of malnutrition in Ugandan children.
The need to breed baboons was determined by the requirement for a regular supply of infant baboons for the development of animal models of protein-energy malnutrition in children. This latter aspect of the work has been described (Coward and Whitehead, 1972) and the successful development of a baboon model for kwashiorkor, a form of malnutrition frequently seen in Uganda, indicated that it would be desirable to continue a programme of breeding baboons in Britain. In addition it must be recognized that it is likely that the supply of primates from the wild will become increasingly difficult and costly (Hartley, 1972). In the long term the only satisfactory way to ensure regular supplies of the animals for experimental purposes, may be to breed them in captivity.
The present work describes the establishment of successful breeding colonies in two quite distinct environments, in Uganda, where baboons are indigenous, and in Britain.

THE UGANDA BABOON COLONY (Figure 1)

An existing colony of 11 females and 5 males, the latter including two adults, was acquired from a zoologist who was leaving Uganda. The fact that the colony had existed for some time and that a hierarchy had been established before the change of ownership meant that unnecessary fighting and disturbance of the breeding programme was minimized, particularly after the removal of one of the adult males. This is an important factor in obtaining rapid results from a breeding colony.
The animals were housed in 4 newly-built pens which were constructed of 3.4 cm steel tubing welded to 5 cm diameter steel mesh; the dimensions of each were approximately 5 m long x 2.5 m wide x 3 m high. Each pen gave access to the next through a trap-door operated remotely from the pen front. The floors were of polished concrete and drainage was into gulleys around the entire periphery. Mains water pressure was very low at first, so an overhead water storage tank was installed which was filled by an electrical pump. This provided adequate water pressure for hose use as well as for a 'Tego' detergent dispenser (T. Goldschmidt Ltd, Harrow, Middlesex).
One of the pens had a trapping cage 1.5 m x 1 m x 1 m fitted to the front.

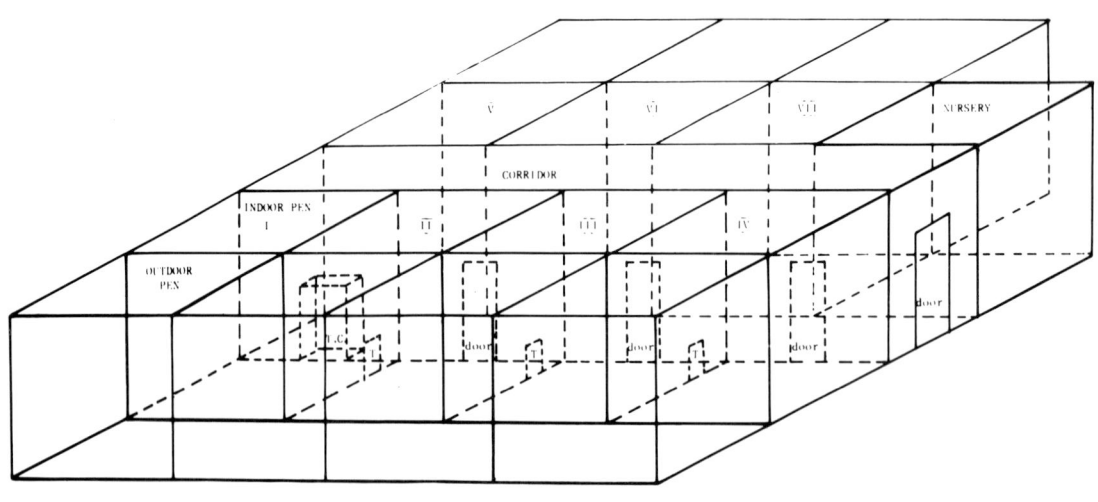

Figure 1. Ugandan baboon accommodation.

This proved to be a weakness in the system and catching individual animals for health screening or for removal of an infant from its mother was a time-consuming procedure. Furthermore, three people were required, two to operate the squeeze-back mechanism and one to inject the animal.

Food was obtained locally and varied depending upon seasonal availability. It included banana, pawpaw, oranges, sugar cane, chickwheat, groundnuts, cassava, sweet potato and elephant grass. Each baboon was given a boiled egg every day. No specially manufactured primate diet was necessary. The food was distributed over the floors of the pens and water was kept in galvanized troughs. These were initially placed at floor level but were later raised to almost roof height when it became obvious that they were being contaminated by excreta. In the higher position this was avoided. When it was found necessary to increase the number of baboons in the Uganda colony, permits were obtained to allow the capture of animals from the wild. Three possible ways of doing this were investigated, using (i) a dart gun, (ii) drugged bait, (iii) a trapping cage. The dart pistol available at the time had a maximum range of 25 m but the accuracy was not good at that range. Drugged bait was tried but was not accepted by the baboons. Furthermore it would have been necessary to track the animal in question until the drug took effect. Dose levels are a problem when using this technique since one would need a drug with a very wide tolerance to avoid killing a small animal if it took the bait. After several unsuccessful attempts at this latter technique, it was abandoned.

A trapping cage was next designed, its dimensions were 123 cm x 77 cm x 100 cm and it was constructed of 5 cm diameter mesh welded to an angle iron framework (Figure 2). It was collapsible for transport and the six sides bolted together at the site of use. The trap-door was held in the raised position by a short bolt which was actuated by a lever attached to the bait by a cord. When the animal entered the cage and picked up the bait, corn-on-the-cob being preferred, this tightened the cord which withdrew the bolt and the heavy trap-door fell into the closed position. A spring loaded catch held the door shut.

It was found that the best trapping area was that used for sleeping by the animals (Figure 3). This exposed them to the traps twice each day, in the

Figure 2. *Cage used for trapping wild baboons.*

morning when leaving their sleeping quarters and in the evening when returning.
  Using this technique on two separate baboon troops, each of 30-40 adults, 25 animals were caught from which six mature females and one mature male were selected for our colony. The rest were marked with dye and released on the same day as captured. The method proved satisfactory and apparently did not frighten the baboons since several were recaptured after being previously

Figure 3. Gorge bordering Lake Albert Flats from which wild baboons were collected.

released as unsuitable. At first large numbers of males were caught but this was later avoided by distributing the traps around the perimeter of the area through which the baboons passed rather than at the entrance; females were then caught in greater numbers.

THE CAMBRIDGE BABOON COLONY (Figure 4)

The Ugandan design had several shortcomings which it was hoped to avoid when designing new baboon accommodation within the confines of an animal unit originally built for small animals at the Dunn Nutritional Unit in Cambridge. Three rooms, each measuring approximately 6.5 m long x 2.2 m wide x 2.8 m high, were set aside for modification as accommodation for the breeding colony. Another room of similar dimensions was available as a baboon nursery and one other was for use as a marmoset room at a later date.

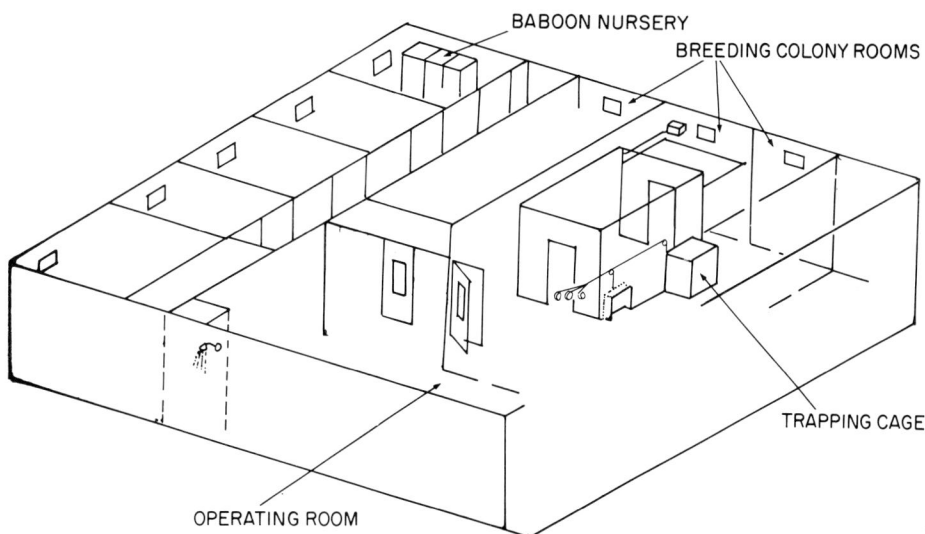

Figure 4. Overall view of Cambridge Animal Unit with detail of baboon accommodation.

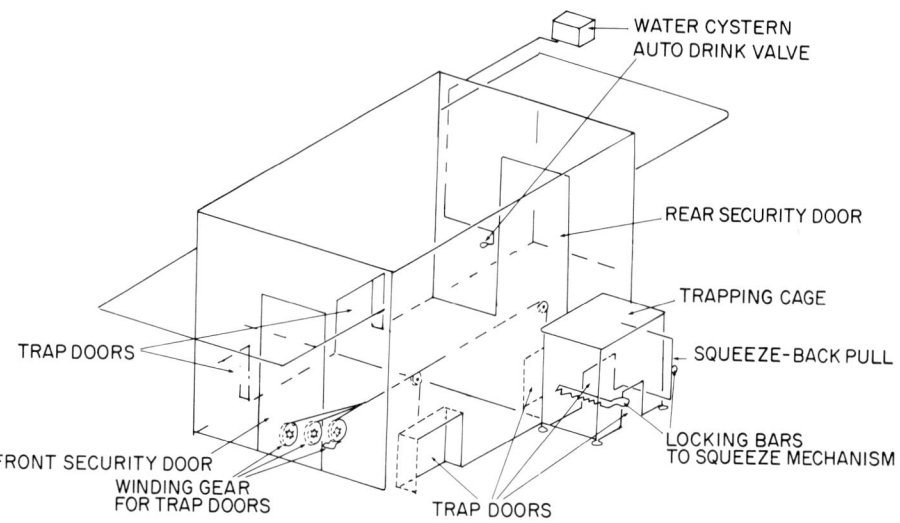

Figure 5. Detail of Cambridge baboon group-cage with attached trapping-cage.

The three pens have been designed and constructed with the accent on security, such that they more than satisfy the requirement of a rabies quarantine area. In fact the baboons in the colony have been effectively maintained in quarantine ever since their arrival.

Each pen is separated from its neighbour by a brick wall equipped with 2 sliding trap-doors (Figure 5), one conventional trap-door and the other associated with the trapping-cage. The front and rear partitions of the pens are constructed in steel mesh with a tubular steel framework like that used in Uganda. There is a padlocked and double bolted mesh door in each partition. The dimensions of each inner pen in which the baboons live are approximately 3 m long x 2.2 m wide x 2.5 m high. A mesh roof protects the lighting and ventilation systems. An overhead mesh barrier also extends beyond both the front and rear of each pen to prevent an escaped animal reaching the water supply cistern to the automatic drinking system and the overhead window which contains an ancilliary extraction fan. A further security door is positioned 2.2 m in front of the pen and beyond that, allowing enough room for a disinfectant foot-bath, is the main entrance door. This contains a wired glass window through which the baboons may be observed without contact or disturbance. This is particularly useful when noting oestrous cycles and matings. These are checked daily and a record kept for each female.

The floors and wall area of all pens are coated with 'PLASIK' (Vigers Stevens and Adams Ltd, Craven Walk, London N16 6BX), a polyester resin with fibre glass which gives a hardwearing and waterproof surface. Stainless steel food hoppers to our design (Forth-Tech Services Ltd, Mayfield, Dalkeith, Midlothian EH22 4AU, Scotland) are screwed to the walls and incorporate an overhead cowling to protect the contents. The bottoms of the food hoppers are perforated for easy cleaning since they are not removable. Water is supplied from two large automatic drinking nipples (Forth-Tech Services Ltd) in each pen. The baboons soon learn to use them. Salt blocks are supplied although to date these do not appear to have been used. The existing heating and ventilation were improved by the addition of an extraction fan in the outer wall of each pen and thermostatically operated fan heaters. With these installed the supply of air is between 13 and 15 changes per hour and the temperature is kept at 20-23° throughout most of the year. The heatwave of the summer of 1976, however, caused the temperature to hover around 27° with a recorded maximum of 32°. All pens operate at negative air pressure. Artificial lighting provides 12 hours of light between 0800 and 2000 hours.

Baboons for the Cambridge colony were obtained through animal suppliers (Shamrock Farms Ltd, Henfield, Sussex). The food provided to the Cambridge colony includes the following: Mazuri primate diet (Cooper Nutrition Products Ltd, Essex, England), groundnuts, brown bread, hard-boiled eggs and oranges.

The cost of feeding one adult baboon in Uganda in 1972 was approximately £0.05/day. At Cambridge in 1976, using our group system, the cost per baboon per day is less than £0.20. The labour required is about 28 man-hours per week.

The main drawback of the baboon accommodation in Uganda had been the positioning and design of the trapping cage. In order to facilitate the capture of a particular animal it was decided to position the trapping cage at the exit of one of the sliding trap-doors between the pens so that in order to pass from one pen to the next the baboons had to go through the trapping cage. The idea was that familiarity would breed contempt and the cage would eventually hold no fears for the animals.

The trapping cage (Figure 6) is designed as a square sided tunnel with the sliding trap-doors at either end remotely operated from the front of the pen from which the animals are removed. Unrequired animals are passed through the trap first, leaving the chosen baboon until last if possible, by judiciously operating the trap-door between the pen and the trapping cage. The exit trap-door from the trapping cage is then closed and the entry trap-door opened. Thus once the animal required passes through the first trap-door the door is closed and the animal trapped. A squeeze-back incorporating a third slide trap-door is situated between the entry and exit trap-doors of the cage and operated by cable from inside the pen in which the trapping cage is situated. Unrequired baboons are cleared from the second pen through additional trap-doors placed in each of the two internal walls between the three pens.

In the Ugandan baboon unit, it had proved necessary for 3 persons to operate

Figure 6. Cambridge trapping-cage.

the trapping cage. In the Cambridge trapping cage the squeeze-back has two arms with right-angled triangular slots cut in them extending forward from the mid point at each side; these engage on two bars at the cage front and act as locking ratchets which interlock successively as the squeeze-back is drawn forward using conventional draw bars fixed to the four corners of the squeeze-back. Once the baboon is confined forward with the squeeze-back and all trap-doors closed, it can be attended to easily. In fact the whole operation could be performed by one person although we have a safety rule in operations of this type that two people must be present.

The trap-door at the cage front was of solid steel plate when first fitted but this somewhat restricted the area of the animal exposed for injection. We have since drilled several holes in the plate to minimize this difficulty; alternatively a square hole could be cut in the trap-door and steel mesh welded over it. This would give even greater accessibility.

To avoid accidentally amputating tails and to reduce the noise level associated with operating the trap-doors, teflon buffers are attached to the bottom edge of all trap-doors leaving a gap of 2 cm between the bottom edge and the floor. All trap-doors are lockable.

The trapping cages have to be protected against interference from within as well as from without, since when the three pens are in normal use each is occupied by baboons. Thus combination locks and chains immobilise the two squeeze-back draw-bars and the ratchet handles are removable after use. The cable used to raise and lower the squeeze-back trap-door is stored behind the squeeze-back inside the cage and out of reach of the baboons. When required for use, it can be removed using a stiff wire hook. When not in use, the cable to the front door of the trapping cage can be unhooked and wound up into the steel tube in which it runs. This front door has a purpose-designed spring catch with a concealed tamper-proof operating lever.

BREEDING OUTPUT

The type of accommodation and the method of husbandry were essentially similar in both the Uganda and the Cambridge colonies, apart from that in Uganda being outdoors where the ambient temperature ranged between $16°$ and $27°$, whereas the colony in Cambridge is indoors at a temperature of between $20°$ and $23°$ in normal British climatic condition.

Table I shows the breeding output over a 4-year period in Uganda. Two abortions and one stillbirth were recorded during this time.

Of the 11 females originally in the colony, one died shortly after a stillbirth and a second one became obese and would not mate. A further female was too old and thus the number of breeding females in the colony for the last three years shown was effectively 8. From the table it can be seen that the maximum birth rate per female per year was 0.75 excluding stillbirths. This is somewhat lower than the figure of 0.9 quoted by Kraemer and Vera Cruz (1972). However their colony numbered 35 producing females and the study was taken over 6 years.

TABLE I

Comparison between yearly birth ratios obtained in Uganda and that obtained during the first year in Cambridge

	Uganda				Cambridge
	1969	1970	1971	1972	1976
Infants born	4	4	4	6	2
Birth ratio (live births/productive female/year)	0.44	0.5	0.5	0.75	0.17

It would be unfair to make a critical comparison between the breeding output during the first year in Cambridge and that in Uganda, since obviously this has been a settling-in period and pecking-orders had to be established. The results were also influenced unfavourably by the death of our first male from an unknown cause three months after arrival. The post mortem revealed nothing except an excessive amount of gas in the gastrointestinal tract. Furthermore, there was a six week time lag before a suitable new male was obtained. A further problem arose from the fact that due to some severe fighting, the colony had to be divided into two groups of 9 and 6 females. This made it difficult to ensure availability of the male to all females during oestrus. In April 1976 another male was introduced so that each group now contains a male. It is hoped to see an increased breeding rate from this time.

Table I also shows the breeding output over the first year in Cambridge. The actual number of females at present in the colony is 15. However, one of these appears to be just reaching maturity and two others would seem to be past their breeding age. The latter two will probably be replaced in the near future. If the 2 abortions and stillbirths were taken into account in computing a potential breeding rate the figure of 0.17 births per productive female per year becomes 0.42. It might be inferred that breeding in a native environment is likely to be the most successful, but a fairer assessment will be possible in the second year of the Cambridge colony. In fact already there are indications of improvement with a further female giving birth in August 1976 and 8 others showing signs of pregnancy.

THE HUSBANDRY OF NEWLY WEANED INFANT BABOONS

Infant baboons are removed from the mother at 10 weeks of age. They are transferred to the nursery which is remote from the main colony; this avoids disturbing the colony should the noise of the infants be audible to them.

The nursery has a lockable outer door with a wired glass window for observation purposes. Within this is a lockable steel framed mesh security door leaving space for a disinfectant foot-bath. The entrance door opens outwards to a corridor. The security door opens in the reverse direction leaving room to move cages through for cleaning. All doors to baboon areas are self-closing. Within the nursery itself, the dimensions of which are 6.5 m x 2.8 m x 2.2 m, there are two tiers of 5 wall-mounted cages 72 cm x 72 cm x 85 cm. They are of the Coid pattern and are easily removable for cleaning, being made of aluminium. The cages are somewhat larger than necessary for infant baboons but they became available and have been utilized in the interest of economy. Ideally, both walls of the nursery would have been equipped with smaller cages, but this would have been in excess of our budgeted expenditure for the baboon accommodation. It is at present only possible to equip one wall with cages. A sliding perspex panel has been introduced at either side of the cages in order that each infant may see its neighbours without interchanging food. The squeeze-back in these cages is unnecessary since the infant baboon soon learns to come out of the cage voluntarily when required for blood sampling and anthropometric measurements. The anaesthetic used during these procedures is 'Ketelar' (Parke, Davis and Co., Pontypool, Monmouthshire, UK) at a dose level of 6 mg/kg. This is sufficient to tranquilize the animal for a period of approximately half an hour, with a fast recovery time.

In Uganda, the infant baboons were weaned on to solids over the first two weeks from removal. The first two baboons born at the Dunn Nutritional Unit,

TABLE II

A comparison between natural baboon milk and the substitute used in this project

	SMA/S-26 + Oil	Baboon Milk[*]
Fat (%)	4.8	4.8
Carbohydrate (%)	7.5	7.5
Protein (%)	1.61	1.6
Energy (Kcals %)	79.6	79.0

[*]Taken from Buss (1968)

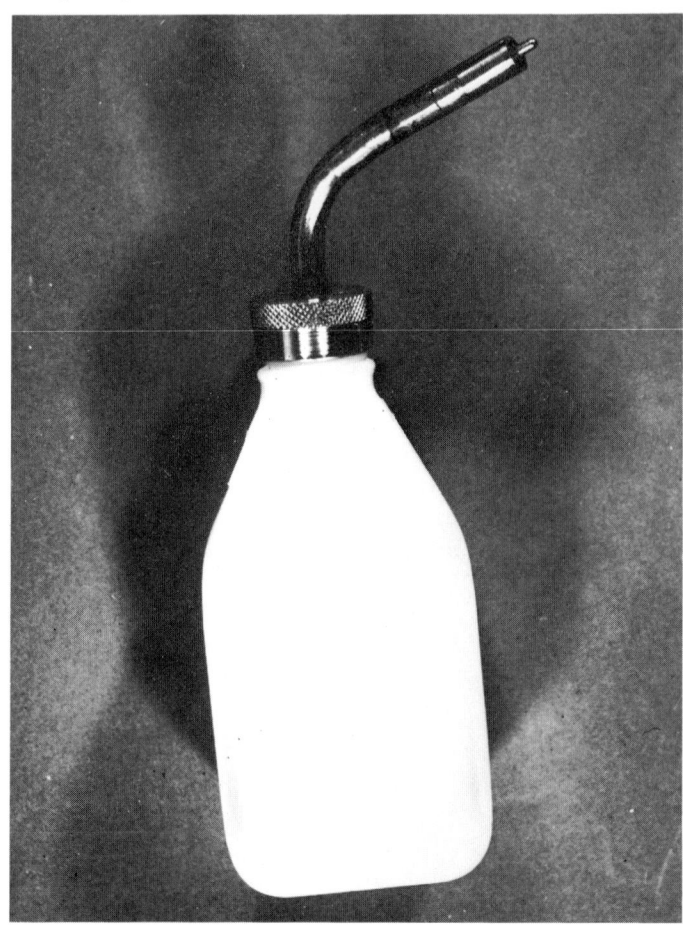

Figure 7. 'Automatic' feeding bottle.

Cambridge, were kept on milk for 6 weeks in order to help establish normal energy intakes. The milk diet was made up according to a modified human formula described by Hummer (1970) using SMA/S-26 (Wyeth Laboratories, Maidenhead, Berks). A mixture consisting of 13.39 g SMA/S-26 + 1.05 g corn oil made up to 100 ml with water resembles the composition of natural baboon milk (Buss, 1968, see Table II).

The first infant baboon was bottle fed five times each day at 0730, 1130, 1400, 1630 and 1900 hours. Two days after removal from its mother, the infant was consuming 250 ml of milk over the five feeds. Its weight at that time was 1.458 kg, its energy intake amounting to 136 Kcals/kg/day. This gradually increased until at 20 days after removal the infant was consuming 450 ml of milk and about 8 g of Farex (Glaxo Farley Foods Ltd, Plymouth, UK) in milk, being equivalent to an energy intake of 240 Kcals/kg/day. The Farex supplement was not taken readily, possibly due to its thick consistency. By the end of the six-week period on the milk, the infant was consuming 500 ml of milk per day which together with 13.5 g of Farex amounted to 243.8 Kcals/kg/day. At this point the animal weighed 1.749 kg.

When the second infant baboon was removed from its mother at 10 weeks of age it was decided to try feeding it in a truly *ad lib* manner by the introduction of automatic feeding. This was achieved by brazing an automatic watering nipple on to the metal screw top of a polythene animal feeding bottle (Figure 7). The bottle was of 500 ml capacity and was filled in the morning and replaced with another full bottle in the evening before the animal technicians left work. At no time was the bottle allowed to empty. By the third day after removal, the animal was consuming 569 ml of the modified milk, equivalent to 308.7 Kcals/kg/day. The volume taken reached 700 ml 25 days after removal, at which point the baboon weighed 1.956 kg equivalent to an energy intake of 285 Kcals/kg/day. The success of the *ad lib* feeding can be judged by the fact that the second baboon gained 90.8 g/week during the first 6 weeks compared to 51.7 g/week for the first (Table III).

There may possibly be some sex related effect here since the first baboon (G/9/1) was female and the second (G/14/1) was male. However, it is unlikely that this would account for twice the rate of growth in the male.

In view of the more consistent growth rate in the second animal, obtained using a considerably less time-consuming feeding procedure, it is intended to continue with this method for subsequent infant baboons, although there is no evidence, other than the slower weight gain, to indicate that the hand fed animal was undernourished. The initial weight loss may have been due to the psychological effects of maternal deprivation.

Regarding neuroses produced in young baboons caged individually, it was found that the second (male) baboon has a more introvert nature than the female. This may have been the result of the female having more contact with her handlers during the bottle feeding period, whereas the male was deprived of this, being fed automatically (Beattie, 1972).

Growth curves, energy intakes and plasma albumin levels obtained in the first two infant baboons born in Cambridge are shown in Figure 8. A decrease in plasma albumin levels, if it had occurred, might have been taken as an indication of an inadequate diet (Whitehead, Coward and Lunn, 1973; Hay, Whitehead and Spicer, 1975).

TABLE III

Comparative weight gain per week and energy intake per kg/day between G/9/1 and G/14/1

Baboon Number	Sex	Weight Gain/Week (g)	Mean Energy Intake (Kcals/kg/day)
G/9/1	F	51.7	196
G/14/1	M	90.8	231

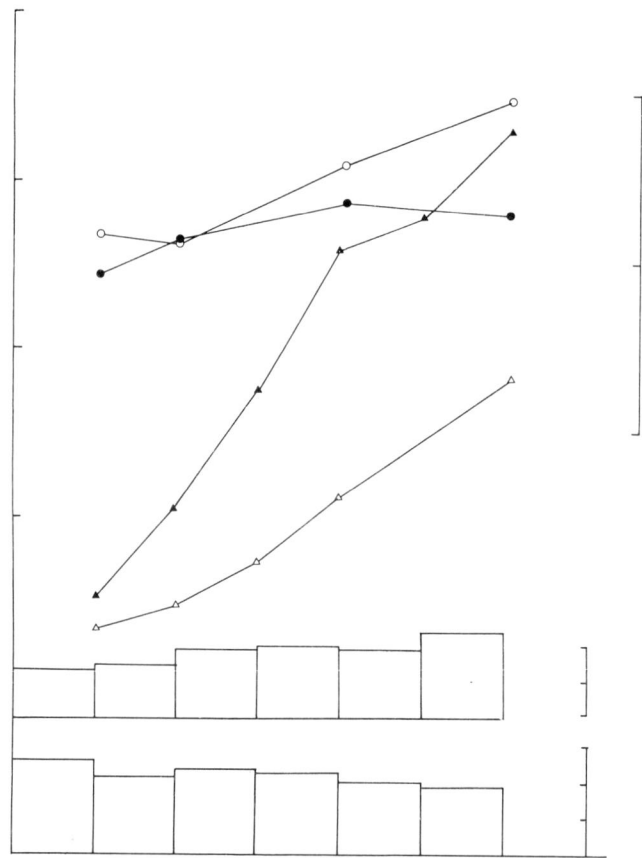

Figure 8. Comparison between female infant baboon G/9/1 and male infant baboon G/14/1 to show body weight, energy intake and plasma albumin levels during the first six weeks after removal from parents.

CONCLUSION

Baboons may be bred in captivity with few problems when housed in a series of interconnecting group pens. This method is satisfactory from both the breeding standpoint and the minimal labour requirement. It reduces the hazard of dealing with baboons and fulfils quarantine regulations. The simple design allows for easy construction and the husbandry and trapping can be done by one person.

ACKNOWLEDGEMENTS

We wish to thank Dr. R.G. Whitehead and Dr. W.A. Coward for their help and encouragement in the preparation of this manuscript. Mr. A.M. Prentice and Mr. B. Porter spent a summer vacation from university in England trapping baboons in Uganda. It was mainly due to their efforts that this project proved a success.

Thanks are also extended to Mr. D.R. Hutt and his staff at the Animal Unit in Cambridge and to Mr. K.R. Symonds for photography and graphics.

REFERENCES

Beattie, I.A. (1972). *In* "Breeding Primates" (W.I.B. Beveridge, ed.), pp. 48-54. S. Karger, Basel.
Buss, D.H. (1968). *J. Nutr.*, 96, 421.
Coward, D.G. and Whitehead, R.G. (1972). *Br. J. Nutr.*, 28, 223.
Hartley, E.G. (1972). *Br. vet J.*, 128, 481.
Hay, R.W., Whitehead, R.G. and Spicer, C.C. (1975). *Lancet*, ii, 427.
Hummer, R.L. (1970). *In* "Feeding and Nutrition of Non-human Primates" (R.S. Harris, ed.), pp. 183-203. Academic Press, New York and London.
Kraemer, D.C. and Vera Cruz, N.C. (1972). *In* "Breeding Primates" (W.I.B. Beveridge, ed.), pp. 42-47. S. Karger, Basel.
Whitehead, R.G., Coward, W.A. and Lunn, P.G. (1973). *Lancet*, i, 63.

THE FINANCIAL IMPLICATIONS OF BREEDING RHESUS MONKEYS FOR BIOMEDICAL RESEARCH

R. KORTE

Primatenzentrum, Münster, Germany.

I want to talk about the breeding of macaques and to limit my topic, on one hand, to the intended breeding with 100 or more female breeding animals; on the other hand, breeding with the aim of a high production of descendants. To make this limitation clear I would like to say that we can talk about breeding when we have more than 5 or 10 animals or when the animals do, slowly but surely, reproduce themselves, for example on an ape rock in a zoo.

This topic may be divided into three parts:
1. The different breeds which are available and especially the methods which are used.
2. The advantages and disadvantages of the methods mentioned above.
3. The planning, construction and management of a large breeding plant with reference to the breeding plant of the PZM, Münster.

It is possible to distinguish between 7 different methods or techniques.

(a) Charles River Breeding Laboratories, Inc. has chosen an island facility. They purchased a small island amongst the Keys in Florida and furnished it with about 1,000 monkeys and are awaiting a good harvest (Figure 1).

(b) A smaller variation to this example (a) would be the system of monkey rock. The rock is surrounded by a water drain and is covered with a latticed railing enclosing the monkeys. This for example is being practised in San Antonio, Texas (Figure 2).

(c) The half-acre corral system is similar. This is being used in Davis, California (Figure 3).

(d) Probably the most common breeding technique is the use of the so-called Corn cribs, as used in Davis, California and in the Texas breeding centre of Hazelton Laboratories (Figure 4).

(e) A further breeding variant is the indoor/outdoor housing system. It is used, for example, by TNO/Holland, in zoological gardens and to a certain extent in the Yamassee breeding centre of Bionetics, USA (Figures 5 and 6).

(f) The extension of system (e) gives complete housing indoors such as now being used by us in Münster (Figure 7).

(g) The last form which can be used is the single-cage indoor housing with the aim of timed pregnancy matings (Figure 8).

After comparing the advantages and disadvantages of the breeding methods mentioned above, it is clear that the single-cage housing for mating, proves to be the best. It is however also clear that this method carries greater expense.

On the other hand, it is also clear, that the monkey rock, the half-acre corral and Charles River's island prove to be the most economical breeding form depending on the colony size and on the geographical location.

Figure 1.

Figure 2.

At this point we must take a few points into consideration comparing the different systems.

(a) The parents should be known in order to have a breeding programme or to assess the breeding results.

(b) The state of health of the animals should always be better than in wild-caught animals. In order to achieve this, health control is necessary.

(c) Provided rules (a) and (b) are followed the best results will be obtained and the costs should be low.

Figure 3.

Figure 4.

It is clear that when the above parameters are considered, only breeding groups of about 8-15 females, together with 1 male, will bring optimum results.

From these groups, the fathers will be known, the groups will not be too large and the rooms will be kept small so that the state of health of each animal can be controlled. For this reason I find that larger operations such as the Charles River Island or the half-acre corral in Davis are not suitable. These operations however, do have the advantage that the expenses for the

Figure 5.

Figure 6.

buildings are kept to a minimum and the heating costs are about zero.

We cannot exclude the fact, however, that in these types of operations infections and pathogenic agents can be carried to the monkeys by birds, rodents, fish, turtles and insects. This is just what we should avoid.

After comparing the advantages and disadvantages it is clear that preference must be given to the kind of breeding unit (1 male with 8-15 females) which avoids such infections to a large extent. The corn cribs have the ideal size and the low costs, but they allow contact with other wild animals (Figures

Figure 7.

Figure 8.

9 and 10).
   The question arises of the lower costs for heating in Texas and California (which is, in turn, dependent on the geographical location) versus the risk of infection through rodents, birds, etc.
   In Europe, the only possibility is to breed macaques in more or less enclosed rooms with appropriate air conditioning. When I say more or less closed, I am thinking of the operation used by TNO in Holland. I believe that when you have gone to the extent of establishing a building it should be

Figure 9.

Figure 10.

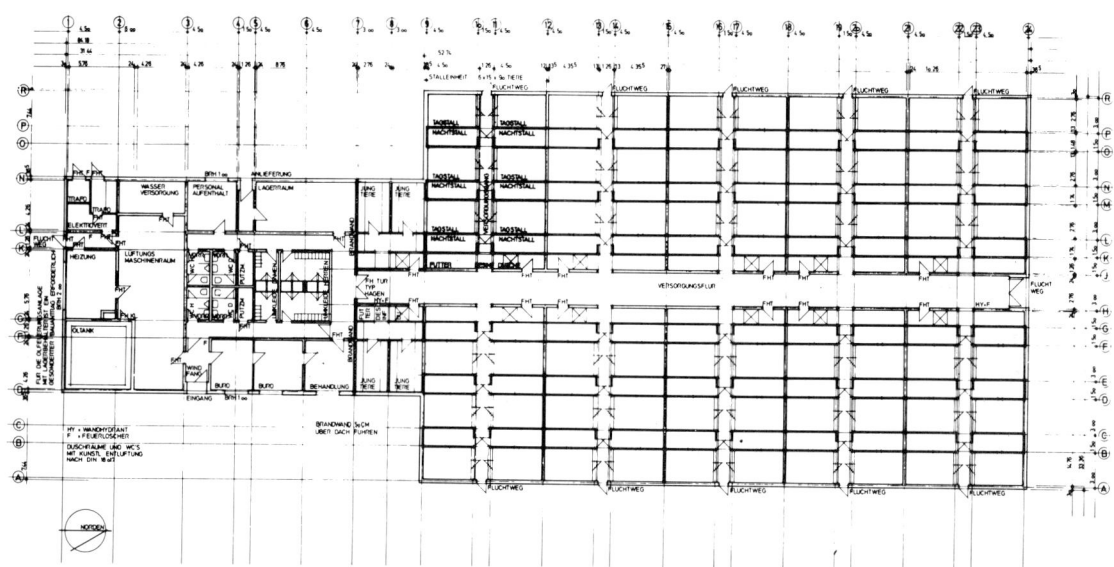

Figure 11. Plan of the breeding colony. On the left side (axes 1-3) the air-conditioning equipment as well as the water-supply and oil-tank are located. Axes 3-7 show offices, storage and social rooms for the animal caretakers, axes 7-9 are the young animal raising areas, axes 9-24 show the 10 breeding units for approximately 1000 breeders.

enclosed. This will have only a small effect on the cost of each monkey.

In consideration we may say that for the northern and middle European areas - in which we want to breed monkeys - when the father and mother are known and when the animals have good health the only breeding method that appears viable is the harem breeding system, 1 male plus 8-15 females in a closed unit.

The facts mentioned above were taken into consideration in establishing our new breeding plant in Münster which will open in 1976 and which will house about 700 breeding females and about 50 male macaques.

During the planning stages we developed various sketches and designs which ranged from the simplest buildings to luxury apartments for primates.

The plan that was finally selected was suitable in the technical sense as well as architecturally. Furthermore it was financially possible (Figures 11, 12, 13 and 14).

The total cost for this project is approximately 2.6 million DM excluding surplus tax.

The various costs are as follows:

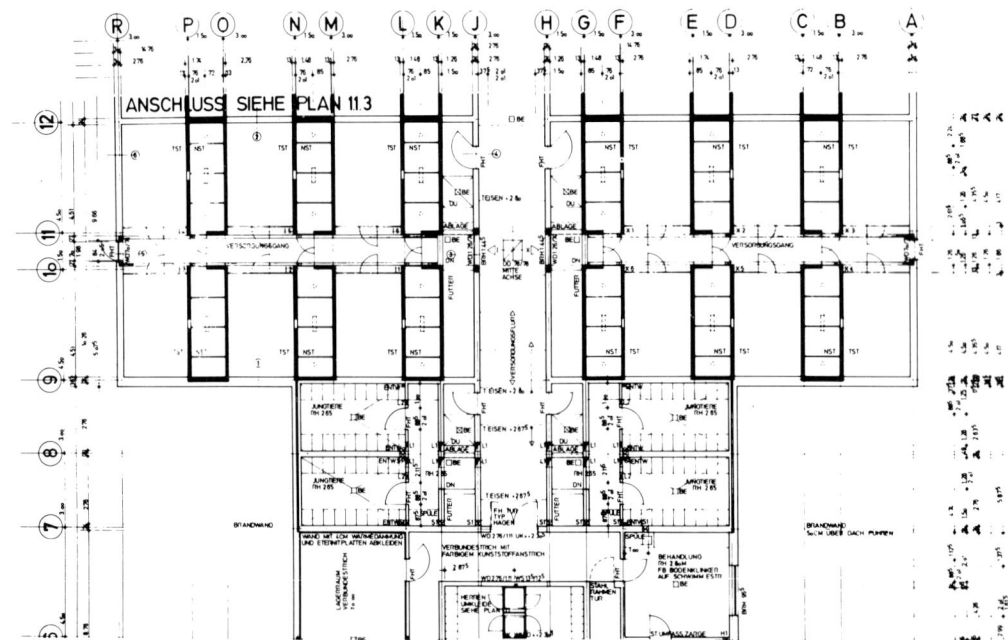

Figure 12. *Detailed plan of axes 6-12. Axes 9-12 show two breeding units, each of which consists of 6 groups with heated and air-conditioned sleeping compartments and unheated runs for the day. Each group of breeders consists of 15 female and 1 male animals.*

1. Cost of the Land

Purchase costs	DM	100,000
Notary costs	DM	1,200
Taxes for real estate register	DM	7,000
Procuring costs	DM	7,500
Contingency	DM	1,300
	DM	117,000

2. Development Costs

Own water supply - included in the sanitary offer		
Electrical supply VEW	DM	11,000
Transformer station	DM	25,000
Sanitary pit - included in the general offer		
Street costs	DM	13,000
Telephone connection	DM	1,000
Contingency	DM	5,000
	DM	55,000

Figure 13.

3. Basic Architectural Costs

The completion of the whole building, including: earthworks, drainage, brick layer-concrete and steel-concrete work, roof construction, roof work, rough-casting, plaster, paint, floor work including joinery and locksmith and the laying of the garden	DM	1,430,000
Complete heating and air conditioning system including the oil tank, boiler, ducting, insulation, etc.	DM	217,000
Complete sanitary installation including water supply, water processing unit with connections for water outlets and indoor equipment	DM	106,000
Complete electrical installation including light fixtures etc.	DM	53,000
	DM	1,806,000

4. Management Equipment and Furnishings

Animal cages and perches	DM	115,000
Single cages for young animal breeding	DM	50,000
Office and laboratory equipment	DM	20,000
Contingency	DM	15,000
	DM	200,000

Figure 14.

5. Additional Building Costs

Architect	DM	68,000
Building management	DM	29,000
Additional costs, architect and building management	DM	10,000
Consulting engineer		
Heating system, air conditioning, etc., for the control of the planning and the take-over of the plan	DM	2,000
Accountant	DM	33,000
Controller	DM	5,000
Survey engineer	DM	3,000
Building permission and the take-over	DM	19,000
Contingency	DM	6,000
	DM	175,000

Combination

Total 1	DM	117,000
Total 2	DM	55,000
Total 3	DM	1,806,000
Total 4	DM	200,000
Total 5	DM	175,000
Total	DM	2,353,000
Animal Purchases (approximately)	DM	330,000
	DM	2,683,000

All operating expenses for the year 1976 are approximately DM 350,000 and for the year 1977 approximately DM 950,000 so that the financial peak at the end of the year should be approximately DM 3.9 million.

The debits made up out of the interest and payments are as follows for the next few years:

DM 1.5 million KfW-M 1 - 7.5% interest, 98% payment, duration 2+8 years, payments twice a year, the first payment on the 30th June, 1978.

1976 Interest/Payment		1977 Interest/Payment		1978 Interest/Payment		
56	-	113	-	109	188	TDM 100

DM 1.0 million payment credit, 8.375% interest, 100% payment, bounded on fixed terms in 6 same yearly rates starting on the 30th June, 1978.

1976 Interest/Payment		1977 Interest/Payment		1978 Interest/Payment		
42	-	84	-	76	167	TDM 100

DM 1.4 million KK-credit, interest D +3.5 until further notice, at the time 7% payments twice a year TDM 100, starting on 31st December, 1986.

1976 Interest/Payment		1977 Interest/Payment		1978 Interest/Payment		
49	-	98	-	98	-	TDM 100

Whole Interest/Payment

| 147 | - | 295 | | 283 | 355 | TDM 100 |

Having studied these figures it is easy to see that with a production of about 300 animals per year approximately DM 1,000 interest falls on each animal. This sum is the highest cost factor in the sale price of the monkeys.

The following calculation details the prices for each monkey:

With a total cost of DM 1,100,000 in the year of 1978, with a production of 335 animals, the cost for a single monkey equals DM 3,284.

This calculation assumes that 50% youngsters in relation to the females would be sold at the age of 1 year.

The pattern mentioned above shows clearly how the prices may vary depending on breeding results:

Breeding Result	Number	Price/DM
50%	335	3,284
60%	402	2,736
70%	469	2,345
80%	536	2,052

Under these circumstances and at the rate of interest shown it is seen that with a high breeding success and with interest-free loans a production price of DM 2,000 can be achieved per monkey. However, it is clear that each monkey of approximately 1 year old will cost DM 3,000 to DM 3,500. Therefore it seems that the breeding of monkeys is more of an economical problem than a biological or technical one.

REPRODUCTION IN A CLOSED COLONY OF *MACACA ARCTOIDES*

J. TROLLOPE

*Department of Growth and Development,
Institute of Child Health, University of London.*

INTRODUCTION

Although great advances have been made in breeding non-human primates in a laboratory environment, the major problem with breeding old world monkeys, such as the genus *Macaca*, seems to be the time (and therefore cost) of producing a usable animal from imported breeding stock, or for captive-born stock to reach sexual maturity.

The survey conducted by Eckstein and Kelly (1966) disposed of the former belief that the conception rates of rhesus monkeys were inferior to those of laboratory rodents. Their findings also emphasised the importance of developing a breeding colony free of experimental interference, giving data to prove the superior performance of colonies with little experimental interruption.

Another factor, which possibly has not been given enough emphasis, is the ruthless selection for breeding potential and health practised by the managers of successful colonies of rodents and other conventional species. Although the managers of large and highly successful breeding colonies of primates practise selection, it is less rigorous than selection in rodent colonies. For example Valerio et al. (1969) consider a female infertile only after 10 to 12 matings without conception. The arguments against selection are obviously (i) the cost and (ii) the potential health hazards even after a quarantine period, of replccing rejected animals in a breeding colony (Short, 1968).

Small invividual colonies, such as the one described in this paper, will be replaced in time by primate centres, as established in the USA and other countries, or breeding programmes founded by large institutes and commercial organisations. However the results obtained without initial stock selection, or replacement and with restrictions on breeding practice, indicate the reproductive potential that old world monkeys of the genus *Macaca* have in restricted laboratory environment with mother-reared young.

ORIGINAL STOCK AND ENVIRONMENT

Two male and 6 female stump-tailed macaques were purchased from a UK importer during October 1966. They were estimated to be in the age range of 6-8 years, as judged by dentition, colouring of the face and genital area, and body weight. We had some difficulty in obtaining adult animals, and no selection for health or breeding potential was possible.

Two of the females were pregnant on arrival, and received the same screening and treatment for pathogens as the other animals. These pregnancies resulted in one full-term still-birth, and a live birth successfully reared.

Three of the original animals were disposed of; two of these (a male and female) are apparently infertile, and the third was a female who had a behavioural problem on arrival, which developed into self-biting.

Since reception the original animals and their progeny have been kept as a closed colony in total isolation from other animals in a room 7.3 x 5.8 x 2.3 m high. Large windows give natural daylight and artificial light is provided from 0730 to 1930 hours. The room temperature is 21 ± 2°. There is no humidity control.

The cages are interconnecting blocks of welded mesh which can provide cages of various sizes by means of removable sides. This system provides visual contact for all the monkeys and limited tactile contact for neighbours. The overall size of the cage block is 460 x 183 cm high. This block is connected to four larger individual cages by sliding doors. The sizes of these cages are 183 x 91.5 cm, 183 x 122 cm, and two of 183 x 183 cm. All four are 183 cm high. In practice one mother with one or two young has a minimum cage size of 183 x 91.5 x 183 cm high. A single adult animal has a cage of 91.5 x 91.5 x 183 cm high. The females with young have two metal resting platforms in their cage, and the single animals one.

THE PRIMARY STUDY

The animals were obtained for an ethological observation study of mother-infant interactions, on a fixed time-table from birth, including the effect of a sibling on this relationship. This meant that the usual intrusive techniques for ascertaining menstrual bleeding and ovulation could not be used. Mating was also restricted to control the age of the older sib at which the next birth occurred, and the young were not removed from their mother until two years or later.

NUTRITION

The main food item is pelleted 'Oxoid' breeding diet, containing 20.4% crude protein. Individual animals are fed 450-520 g per day. Pregnant animals and mothers with young are given pellets *ad libitum*. This diet is supplemented with oranges, carrots, bananas, boiled eggs, cabbage and bread. The mothers with young are given enough food supplements to ensure the young are not deprived by the mother. Water is given *ad libitum*.

MENSTRUAL CYCLES

Visual inspection of the vagina to detect menstrual bleeding is carried out daily. Small amounts of favourite food are placed in feeding hoppers 1.3 m high and when the females climb on to these to eat, visual inspection of the vagina is carried out.

It would appear that overt menstrual bleeding is less in *M. arctoides* than that found in some congeners. In some species, such as *M. nemestrina*, genital area swelling and colour indicates the ovulatory period during a menstrual cycle (White et al., 1973). In our *M. arctoides*, a slight increase in colour and swelling of the ano-genital area, dorsal and lateral to the tail root and sometimes including the vulva, has been noted in females showing periods of menstrual bleeding. However, this does not vary within the cycle, or show any discernible peak. Our estimate of the length of cycles is taken as the mean of 30 cycles from periods of regular menstrual bleeding from six females.

This was 29.4 days, with a mean duration of flow of 3.7 days (Trollope and Blurton Jones, 1975).

MATING

The mating period was limited to one hour per day, although the two females which were rejected as infertile were occasionally left with a male overnight. The mating method adopted was females to males at days 11-12 of the menstrual cycle, counting the first day of bleeding as day one of the cycle, for the first 8 months of mating. Afterwards this was extended to days 10-14 when possible. Once a female had given birth all matings were restricted to avoid clashing with the observation schedule. Remating the mother to obtain a second baby for the study of a sibling relationship became a problem. This was solved by using sliding doors to move the adult male, so that one empty cage was between him and the female with the infant, then separating the mother and infant by moving the mother into the empty cage, then removing the sliding door so that male and female were caged together. The young animal then had visual and tactile contact (through the wire mesh) with its mother during the mating time, which was never more than one hour. All these mother-with-young matings have taken place without obvious upset to the mother or offspring; the youngest infant to be separated has been four months of age.

Wounding of females during the first 9-10 months of matings occurred fairly frequently. After this period wounding became rare and the only incident during the last four years was a very minor wound inflicted by a laboratory-born 6-year-old male on his female partner. This was the only occasion on which a laboratory-born male has wounded a female.

As can be seen in Table I, the females were mated by exposure to the male for not more than one hour a day, with the one exception of female 8, whose first pregnancy was obtained by caging with a laboratory-born male for 189 days. The average number of matings to obtain a pregnancy in four imported females was 7.5, 11.2, 15.5 and 19 respectively. However, some of these matings were carried out without prior menstrual bleeding indicating a possible ovulatory cycle and often when bleeding was demonstrated the female could not be mated due to study restrictions. The best results, from female 1 (5 pregnancies) and female 2 (7 pregnancies - 6 conceived in the laboratory) were obtained because these females demonstrated overt menstrual bleeding more often than the other females. The poor results from the laboratory-born females were due to the almost total absence of observable menstrual bleeding in these two females. Other laboratory-born females could not be mated, although sexually mature, because of restrictions on space available in one room. Most of the laboratory-born males had to be disposed with, to prevent overcrowding, and the three males retained proved to be fertile.

Two of these males demonstrated temporary sexual incompetence, when first mated to imported and laboratory-born females. This incompetence consisted of an apparent inability to complete normal copulation. However this failure only occurred while older dominant males were in visual contact. When visual contact was broken, both young males demonstrated normal copulations. This transformation was immediate, normal copulation occurring within seconds of the dominant male being visually isolated. A similar incidence of a dominant animal affecting the reproductive behaviour of a subordinate animal has been described by Lang (1972). The species concerned was *Cercocebus atterimus*.

TABLE I

Results from Five Imported and Two Laboratory-Born Females October 1966 to February 1976

Female Number	Origin	Date first Mated	Young from Females POA*	Young Conceived in Lab.	Exposure to Males Total Days/Hr***	Age (Yrs) of Previous Young when Remated	Total Time (Yrs) Available for Breeding to Date
1	Import Oct 1966	Jan 1967	-	5	56	1.58 1.58 1.00 0.66	7.09
2	Import Oct 1966	Mar 1967	1 still-birth	6	57	0.33 0.92 0.58 0.83 0.67	6.90
4	Import Oct 1966	Dec 1967	1	3	57	1.00 0.83 1.08	5.33
5	Import Oct 1966	Oct 1967	-	2	31	1.08	6.42
6**	Import Oct 1966	Mar 1968	-	1	10		2.58
8	Born in lab.	Nov 1970	-	2	Caged with male 85 + 189 days	1.00	5.08
9	Born in lab.	Jan 1974	-	1	34		1.0

*POA, pregnant on arrival; **Animal disposed of, unsatisfactory breeder; ***Exposure to male one days/hr.

PREGNANCY

From a total of 22 pregnancies, 20 conceived in the laboratory, 21 have proceeded to term without incident. With the scattered mating dates, exact gestation periods have not been generally available. We have since obtained five pregnancies from one-hour exposure to a male on one day only. The mean of these five exactly known gestations is 178 days (Trollope and Blurton Jones, 1975). Macdonald (1971) records that the gestation length of 10 *M. arctoides* live births was 177.5 ± 2.1 days.

PARTURITION

Laboratory-conceived births have occurred in every month of the year, except May (Figure 1). The 21 successful births, 13 males and 8 females, have included 2 second-generation births from laboratory-born parents, one from an imported father and laboratory-born daughter, one from an imported mother and

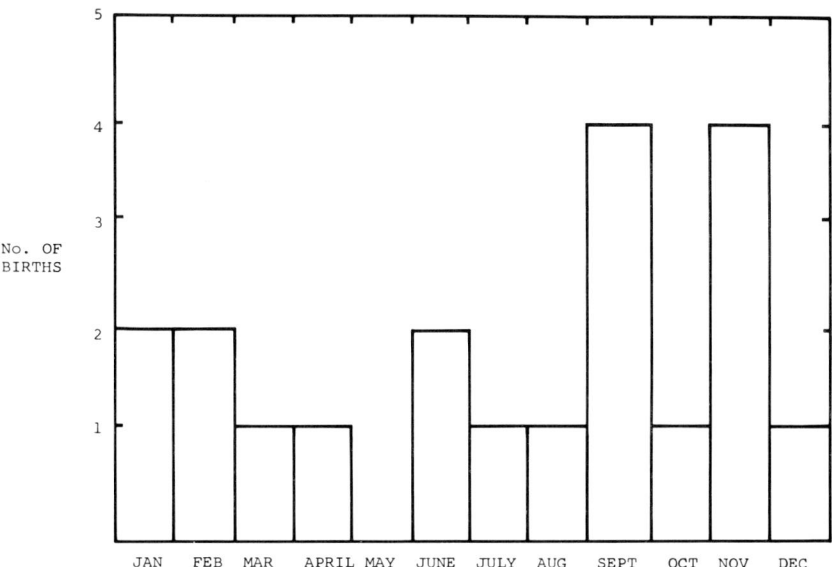

Figure 1. *Twenty laboratory-conceived births from January 1967- February 1975.*

laboratory-born son, and one from an imported female mated to an unrelated laboratory-born male (Figure 2).

With one exception, all the births have occurred at night, the neonates being clean, dry and clinging to the mother the following morning. The exception was the fifth consecutive laboratory birth for an imported female (female 1). On the day of birth this female had a slight vaginal bleed at 1030 hours, at 1400 hours she began to manipulate her genital area, while sitting on a platform. She was tolerant of the attentions made by her previous young, a year-old male, holding him while he was in ventral-ventral contact. Labour was judged to have commenced at 1555 hours, the manipulation of the genital area continuing, and the female alternately lying down and standing on the platform. The neonate's head emerged at 1630 hours, the female reaching back between her legs, whilst standing, and pulling at the head with her left hand. When the neonate was fully emerged at 1655 hours, the female began biting, manipulating and licking at the membrane covering the head. The neonate vocalized and moved her arms vigorously at this stage. The female then climbed down from the platform carrying the neonate, and began to walk the floor. The observers then left the room, in case their further presence worried the mother. The next morning all was well, the neonate dry and on the nipple, and the placenta apparently eaten. Only on one occasion has a placenta been found the morning following the birth, and this was partly eaten. Gouzoules (1974) has described a parturition in a captive group of *M. arctoides*, and the responses of group members. These largely consisted of group members approaching and visually inspecting the female and her neonate, also giving vocalizations. Tactile contact consisting of the female being groomed by a male and an elder sibling touched the neonate. Our female was groomed prior to the birth (through the wire mesh) by females in adjoining cages; at the time of birth they became very interested, moving to look at the neonate, and vocalizing. The older sibling ran around the cage, approaching and sniffing at the neonate, although tactile contact was not made.

248   J. Trollope

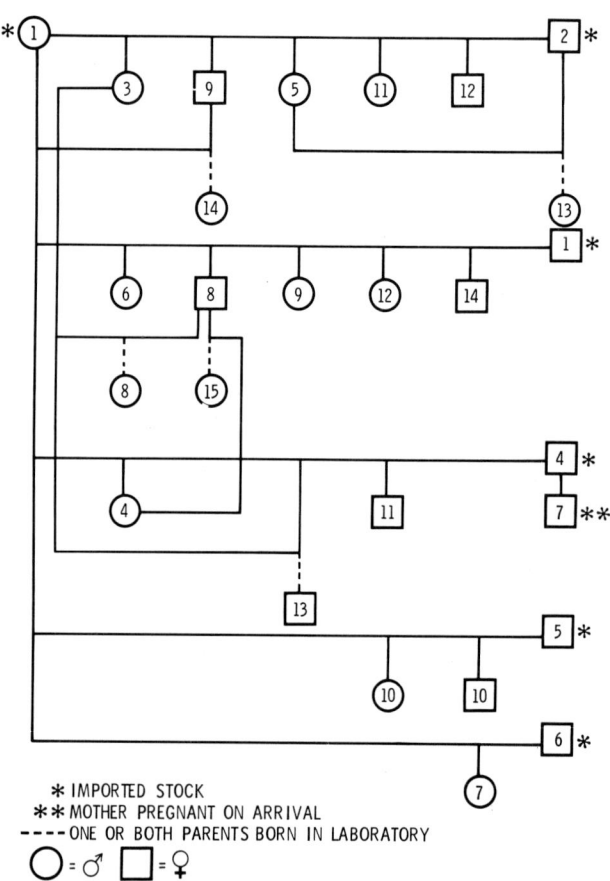

Figure 2. Parent-offspring relationships, initial stock one male and five females.

The females in our colony are usually stiff and reluctant to climb the first day or two post partum. Post-parturition bleeding has lasted 2-7 days, and is variable in amount of flow.

LACTATION AND REARING

Lactation for seven young has lasted 6, 7, 8, 8, 10, 14 and 15 months (as judged by milk seen when the mother expressed it manually, or when the baby paused during feeding). Young have been accepted on the nipple longer than this, until the next birth, when rejection from the nipple is abrupt and final. However the mother will still hold the older sibling and they will huddle together for sleeping periods, with the neonate or infant enclosed between the mother and older sibling.

Female 1 has reared two young after one of her nipples was bitten in a fight through the wire with another female. This nipple had most of the tissue removed and became atrophied and useless.

No problems have been apparent with these triad groups, and the only mortality from birth until permanent separation from the mother at two years or later has been one infant death at six months. The mother of this infant (female 6) was slow to recover from the stresses of importation, and on reception had a behavioural problem, which developed into self-biting.

MENARCHE AND SEXUAL MATURATION

The age of the first recorded menstrual bleeding from one laboratory-born female and five females conceived and born in the laboratory, has decreased through the period of captivity from 835 days to 602 days (Trollope and Blurton Jones, 1975). Decrease of the menarche age was observed by Van Wagenen (1972) in the large established Yale colony of *M. mulatta*, the average age for the onset of menarche in this colony being 624 days, which was 3 months earlier than the record for the first 10 years.

Of the 2 births from laboratory-born parents, the age of the male parent at the time of the first birth was 3.75 years and the female 3.45 years (Trollope and Blurton Jones, 1972). As the parents were caged together for 189 days prior to the birth, we used 170 days as the approximate gestation period for *M. arctoides*, thereby calculating the onset of sexual maturity at 3.25 years for the male and 3.0 years for the female parent. Maple et al. (1973) record the onset of sexual maturation in two laboratory-born *M. mulatta* pairs as 3.30 years for the males and 3.11 and 3.13 years for the females respectively. This indicates that the onset of sexual maturity for laboratory-born macaques can be earlier than the generally accepted 3.5 for females and 4.5 for the males (Napier and Napier, 1967).

DISCUSSION

Where future breeding stock is required and social experience for young animals necessary, combined with known gestation periods and other data in a laboratory single-cage system, we think our method of remating the female without permanent removal of her current infant could be adapted, with the use of the usual techniques to ascertain menstrual bleeding etc., to produce realistic breeding results.

This method would certainly be less costly than removal of the neonates soon after birth for hand-rearing, and possibly less costly than early weaning of the infants at 3 months, when it is the usual practice to provide extra care, usually in the form of a special diet (Goosen, 1972).

Another factor is the fecundity record and long breeding life of selected macaque females in a laboratory environment. Van Wagenen (1972) records the histories of three multiparous females in the Yale *M. mulatta* colony. These were 14, 13, and 12 pregnancies respectively. Valerio et al. (1969) had several females producing five births and one monkey had seven births within a five year period. It is the policy at both these laboratories to remove the neonate soon after birth for hand rearing.

There do not appear to be comparative data for multiparity records for large open-air compound groups. However, Goosen (1972) reports a female in a sixth pregnancy, with a breeding method of small harem groups of one male to four females. The best performance of our females imported as adults is female 1, five laboratory-conceived pregnancies, and female 2, seven pregnancies, six conceived in the laboratory. This indicates that with initial selection for breeding potential and replacement of infertile stock, female macaques can have good multiparity records in a laboratory environment, without early weaning of young.

The main advantage of group breeding in a laboratory environment or group breeding in an open-air compound is undoubtedly the reduced cost per animal reared, compared to the single-cage laboratory-breeding methods. Another adventage is the social experience gained by young animals required for future breeding stock. But this factor is dependent of the age of weaning, and post-weaning social contact.

The problems of group breeding seem to arise mainly from social conflict within the group. Blakely et al. (1972) reported wounding of females and losses of infants in *M. nemestrina* laboratory harem groups. The effects of these conflicts were reduced by avoiding the introduction of new group members whenever possible, and removal of the male canines.

With a large open-air group of *M. mulatta*, Bourne and Golarz de Bourne (1975) found that the introduction of new group members, or the reintroduction of former members, lead to conflict and injury. This problem was solved by the setting up of smaller subsidiary breeding colonies, in corn cribs.

It is impossible to assess the effects of social conflict in a breeding group, without comprehensive data on a variety of possible losses from conflict. Such losses might include the time and therefore breeding potential lost from injuries inflicted on group members, the cost of medical care, subsidiary housing etc., and the adverse social effects of group conflict which might lead to reduced breeding potential from interference with copulation etc.

SUMMARY

From a stock of two male and six female *M. arctoides*, 22 offspring, 20 conceived in a laboratory, have been born, and 20 successfully reared. Laboratory conceived young have included two births from laboratory-born parents, one from an imported father and laboratory-born daughter, one from an imported mother and laboratory-born son, and one from an imported female mated to an unrelated laboratory-born male.

Two of the females imported as adults had 5 laboratory-conceived pregnancies, and 7 pregnancies, 6 conceived in the laboratory respectively.

Access between sexes has been limited due to a primary study, and young are not removed from the mother until two years or later.

ACKNOWLEDGEMENTS

I am most grateful to Dr. N.G. Blurton Jones and Professor J.M. Tanner for their continuous support and encouragement. I would also like to thank Mr. C. Orgill for technical assistance and Miss Anne Phillips for typing the manuscript. The work was supported by grants from the Nuffield Foundation and the Science Research Council.

REFERENCES

Blakely, G.A., Morton, W.R. and Smith, O.A. (1972). *In* "Medical Primatology"
    (E.I. Goldsmith and J. Moor-Jankowski, eds), Part I, 61-72 (Karger, Basel).
Bourne, G.H. and Gorlaz de Bourne, M.N. (1975). *In* "The Rhesus Monkey"
    (G.H. Bourne, ed.), Vol. 2, 262-276, Academic Press, London and New York.
Ekstein, P. and Kelly, W.A. (1966). *Symp. Zool. Soc. London.*, 17-91.
Goosen, C. (1972). *Breeding Primates*, 88-91. (Karger, Basel).

Further references to be found on page 312.

KEY LOIS ISLAND PRIMATE BREEDING PROJECT:
A 3-YEAR PROGRESS REPORT

J. PUCAK

*Charles River Breeding Laboratories, Inc.,
251 Ballardvale Street, Wilmington,
Massachusetts 01887 USA.*

INTRODUCTION

During the past three years much valuable information and experience have been obtained and used in the design of new facilities. This resulted in improved animal care and maintenance. Data on feed and water consumption have also been collected and have helped in the development of routine procedures

Figure 1. *Corral showing the smooth interior walls, the rough outside walls and the exposed cyclone fence at the bottom allowing for ventilation.*

Figure 2. Aerial photography showing the overall construction design of the first corral.

to insure their continuous availability. In addition, critical information on reproduction has been accumulated since we have now completed our third year of breeding rhesus monkeys (*Macaca mulatta*) on Key Lois Island. Based on this cumulative information, it is becoming more evident that breeding rhesus monkeys in a free-ranging environment can be a feasible and sound method for insuring the continued supply of these animals to the research community.

FACILITY DESIGN

The initial importation, quarantine protocols and facilities in Wilmington, Massachusetts, as well as the base house support facilities in the Florida Keys, USA and island preparations have been documented by Foster (1975 and 1976) and Pucak and Foster (1975). The first holding and trapping corral on Key Lois Island is a 465 sq m enclosure that measures 15 x 30 m. The framework consists of a 2.4 m high cyclone fence wall to which 1.2 x 2.4 m galvanized steel panels are attached to the inside, approximately 1 m from the ground (Figure 1). A 4.5 m sq by 4.5 m high plywood enclosed shelter was provided within the corral for protection during tropical storms and for the purpose of conveniently trapping animals. At each end of the corral there is a 1.2 x 2.4 x 2.4 m cyclone fence enclosure to which a 13 mm pipe extending from water storage tanks is attached. Four Lixit valves (Systems Engineering, PO Box 2580, 461 Walnut Street, Napa, California 94558 USA) are provided to

Figure 3. *Aerial photograph showing the smooth inner and outer walls of the second corral design.*

offer water *ad libitum* to the animals. This corral design prevents escape of animals that are being held, but allows animals to enter the compound from the outside by climbing over the outside walls if we wish to trap them after their release (Figure 2).

Obviously once the first group of animals was released, the compound could no longer be used as a conditioning area for new groups since prior groups could easily re-enter. Therefore, our second corral design is basically identical to the first with the exception that additional 1.2 x 2.4 m galvanized steel panels are placed on the outside wall thereby preventing outside animals from entering the compound and not allowing animals being conditioned from escaping until we purposely release them (Figure 3).

Figure 4. The first modified design for a feed, water, trapping and protection shelter.

Two corrals of this latter design were constructed and used to condition and trap animals. As additional animals were added to the island population, there was an increased need for feed, water, trapping and protection sites without a need for the previous elaborate compound design. Our initial design for this type of structure included the 4.6 x 4.6 x 4.6 m high plywood enclosed shelter to which 3.0 x 4.6 m cyclone fence enclosures were added on at each end (Figure 4).

Inside the cyclone fence enclosures, water was provided *ad libitum* via the mechanism previously described, and food was provided *ad libitum* in standard hog feeders 61 x 58 x 91 cm. The entrance to this enclosure was in the floor of the protection shelter. This meant that animals would enter from the bottom, move to the porches to feed and drink and could be trapped inside via a sliding door.

At this time, it should be pointed out that all of these facilities were built on the narrow sand ridge which runs along the entire east-west border on the Atlantic side of the island (Figure 5).

Another type of structure was designed for several purposes: to reduce the concentration of animals along this sand ridge; to utilize the vast space in the center of the island; to provide food and water sites to the various troops; and to minimize aggression. Initially, a 1.8 m wide path approximately 274 m long was cut through the mangrove swamp across the island. A 1.5 m wide floating walkway was constructed over this path. It consisted of

Figure 5. Aerial photograph of Key Lois Island which shows the sand ridge (clear areas) which runs along its Atlantic Ocean border.

aluminium I beams attached to Styrofoam blocks. Specially prepared plywood (High Density Overlaid Surface Plywood, Georgia Pacific, 900 Southwest 5th Avenue, Portland, Oregon 97204 USA) was bolted to the aluminium beams to make a smooth surface. This made walking across the island easy and allowed a small tractor and trailer to cross the island rapidly. At a point along the main walkway approximately 167 m into the center of the island, a 2.4 m x 15.2 m extension was built perpendicular to the main walkway. At the end of this extension a 7.6 x 15.2 m floating platform was built. On this floating platform a 7.3 x 14.6 x 2.4 m cyclone fence enclosure including the roof was constructed. At each end 1.2 x 2.4 m openings fitted with spring-loaded remote control sliding doors are used for entrances. Feed and water facilities are located in the center of these enclosures (Figure 6). Once animals enter, a technician can release remotely the doors, thereby allowing us to capture small groups of animals for harvesting from the island or for testing during the routine monitoring program.

The original water reservoir system, described by Foster (1975), has been expanded to 21 rectangular fiberglass-coated aluminium water tanks holding 985 liters each, and two 4,542 liter fiberglass tanks, giving us a total capacity of 29,750 liters of water on the island. In addition, one 4,542 liter tank has been placed at the base house in case water supply to the base house is interrupted. The island tanks are all interconnected and water is pumped to these tanks using electric pumps operated by gasoline-powered generators.

Figure 6. The floating walkway in the foreground and the cyclone fence enclosure resting on the floating platform.

This water reservoir represents an approximate 20-day supply at current usage (Figure 7). Water is chlorinated (5 ppm) and taken to the island via 1,249 liter tanks that are fixed to the inside of shallow draft fiberglass boats. Each boat is 7 m long with a 2.7 m beam and each is powered by twin 115 hp outboard engines.

Standard commercial monkey feed is stored at the mainland base house facility in a 3.1 x 7.6 m walk-in cooler and is taken to the island on a daily basis. The exterior of all bags are sprayed with a 5% chlorine solution prior to being taken to the island. In case of inclement weather, a 1-2-day supply of feed is kept in a storage shed on the island and older feed is given first to minimize the possibility of nutritional deterioration.

A small laboratory has always been maintained at the mainland base house. However, significant renovations are now completed enabling all the testing involved in the routine colony monitoring program to be performed. This will include bacteriology, hematology and parasitology. Only the testing for herpes B-virus antibody will be done at the main Charles River Laboratory in Wilmington, Massachusetts, and it will remain the reference laboratory for all procedures and provide assistance during peak testing periods.

The original corral on the island is being renovated to facilitate and improve the colony monitoring program and to have a holding conditioning area for animals prior to being shipped. The smooth galvanized metal panels are

Figure 7. Six of the twenty-one fiberglass-coated aluminium water tanks and the two large fiberglass tanks all inter-connected for water transfer.

being removed exposing the cyclone fence. A roof consisting of aluminium I beams and cross pieces form the framework. Corrugated green fiberglass panels 0.7 x 2.4 m are placed in the framework and the entire roof is then covered with 3.75 cm diamond mesh cyclone fence. Inside this structure approximately 200 stainless steel cages mounted on aluminium framework will be used to hold the animals (Figure 8).

MONITORING

Animals introduced to the island were free of tuberculosis, herpes B-virus antibodies, salmonella and shigella (Orcutt et al., 1976; Foster, 1976; Pucak and Foster, 1975). The elimination of the shigella carrier state was achieved by the use of trimethoprim-sulfamethoxazole (Pucak and Foster, 1974). During the period included in this report, monitoring of animals on the island includes some 1,063 serum samples from 566 monkeys tested for the presence of herpes B-virus antibody. Also, approximately 4,700 direct rectal and fecal enrichment cultures from 302 monkeys have been tested for the presence of enteric pathogens, including salmonella and shigella. Monitoring for the presence of tuberculosis by the palpebral intradermal injection of 0.1 ml of undiluted mammalian tuberculin (provided by United States Department of

Figure 8. Interior of the holding compound with individual cages.

Agriculture, New England Area (1) Office, 424 Trapelo Road, Waltham, Massachusetts 02154 USA) was done at the time animals were handled for the above testing procedures.

In addition, freshly voided fecal samples have been routinely collected at random and tested for the presence of shigella and salmonella. Voided feces were also collected and fecal flotations done to determine the presence or absence of endoparasites.

With the development of a holding facility on the island, expanded laboratory facilities and the new feeding/trapping station design, we will have the ability to trap small groups of animals, individually cage them, and significantly expand the monitoring program to permit the individual monitoring of each animal at least annually.

BEHAVIOR AND REPRODUCTION

All breeding animals are permanently identified with a chest tattoo. Animals are identified by technicians, and the date and location on the island where the animal is observed are recorded in a master inventory file on a daily basis. This allows us to document over a period of time the presence and condition of an animal and has provided preliminary behavioral information. From these records and from other observations by a primate behaviorist, we have evidence that currently there are six organized troops of animals. Four of these are breeding troops and contain adult males and females, young adult

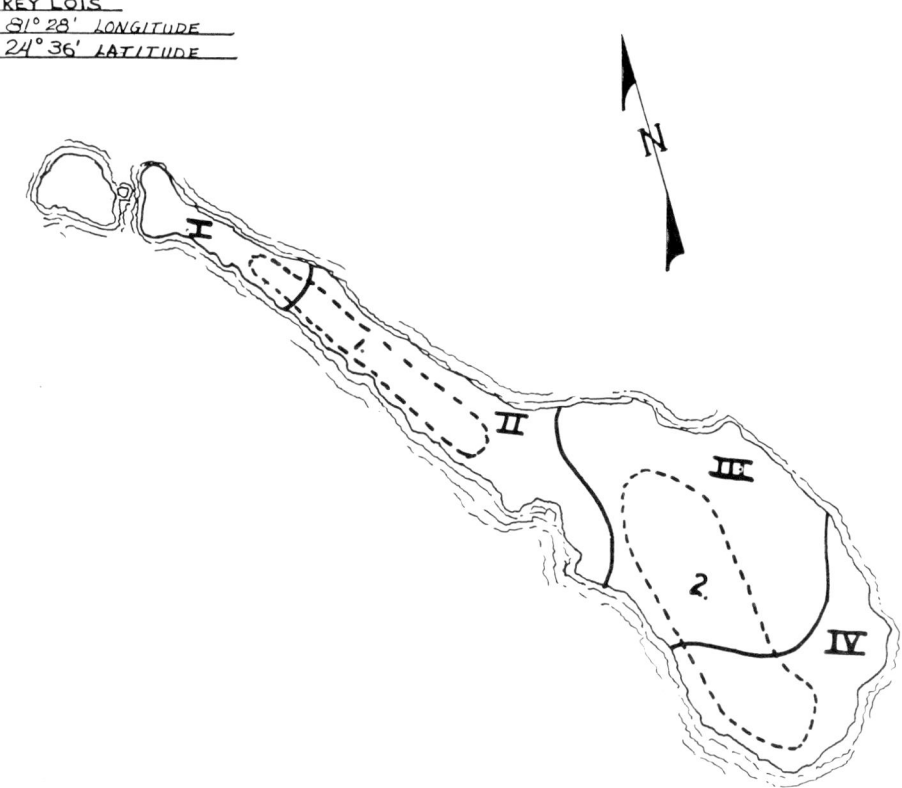

Figure 9. Map of Key Lois Island showing the current troop and subgroup territories. Roman numerals indicate troop territories, Arabic numbers indicate subgroups or bands.

females, juvenile or weanling males and females, and infants. The only young adult males seen in these troops are part of the social hierarchy. However, there are two other subgroups or bands which vary in size from twenty to thirty animals and are comprised of young adult males.

The island appears to be broken primarily into four troop territories (Figure 9).

Troop I

Troop I is the largest and dominant troop on the island. It includes 400 to 425 adults and approximately 175 to 200 newborns to 1½-year old animals. This troop's territory is approximately eight hectares and is at the narrow end of the island. It is the most established troop having as its leader one of the original large males put on the island. Five to seven subleaders assist in the maintenance of this group. Their territory, including feed and water areas, are well defined and defended.

Troop II

Troop II consists of approximately 225 adults and 125 to 150 newborns to 1½-year-old animals. Their territory includes approximately eight hectares, but this group is less established and is a bit more transient in its

TABLE I

Rhesus monkey breeding data, Key Lois Island, Florida, USA, 1974-1976

Year	Births	Estimated adult female population	Breeding efficiency % reproduction	Infant losses up to six months of age
1974	34	45-50	75-68	0
1975	162	225-250	72-64	3
1976	344	450-500	75-68	17

movements to feed and water areas. Unlike Troop I, it is, at this time, less likely to defend a feeding area vigorously. When threatened by other groups, it tends to move to one of the other feeding sites in its territory. It has also adjusted its feeding times to minimize confrontations with the other troops. It is led by a dominant male and two to five subdominant leaders. The social structure of this group is not as tightly bound as Troop I, and females, especially young adults, frequently have interchange with Troops III and IV.

Troop III

Troop III is a poorly organized troop that roams over an area of approximately twelve hectares. It includes approximately 275 animals with a large portion of young adult females. There are also forty to fifty animals less than two years old. The leadership of this group has not been firmly established, and currently there is competition between four large males. Fragmentation of this group into subgroups with such leadership is frequently observed during feeding times. During the major part of the day these small groups do congregate and move about their territory as a unit. As in Troop II, many of the young adult females move to different feeding sites and freely mingle with the various subgroups.

Troop IV

Although Troop IV comprises the greatest number of multiparous females, there does not seem to be a dominant male leading this troop. Several different males have been observed to lead them through their territory, but no one of them is consistently dominant. This group has approximately 150 adults and 100 to 125 animals less than two years old. Several females in this group are now nursing their third infant. They do move as a unit but are easily displaced at a feeding station and tend to move to another rather than defend the one they are at.

There are also two subgroups each comprising approximately 25 young adult males. These two groups move as units, one tending to stay in the territories of Troops I and II. The second moves through the territory of Troop III. These animals are frequently displaced from feeding/watering sites and spend much of their time moving to the various feed and water stations.

The intermittent introduction of new groups of animals has made the evaluation of these data difficult. However, in April 1976, the final group of animals was introduced to Key Lois Island. We do anticipate some additional reorganization in all troops since most of the females were placed on the island when they were $2\frac{1}{2}$ to $3\frac{1}{2}$ years old and are now reaching ages when they will begin to join established breeding groups. Also, young adult males will begin to assert themselves during the breeding season.

We are now approaching the end of our third year of production on the island, and the results are very encouraging (Table I).

We are ahead of estimated production goals and hope that by 1978-1979 an annual production of 800-900 animals will be realized. By that time all the females introduced as juvenile animals will be of breeding age, and males that were introduced as young adults will be mature. In addition, because production has been better than expected, we can begin to cull and replace non-breeding or old breeding females. This will ensure a scheduled replacement of animals in the breeding colony. In the fall of 1976 we will also begin to collect and make available to the research community some 50 to 100 males which will be approximately two years old.

During the trapping of animals for routine monitoring, infant males and females that are with their mother will be permanently identified at this time. This will enable us to document the maternal lineage of future breeders and make selections for breeder replacement.

CONCLUSIONS

Facilities

As project requirements changed, four different corral designs were developed for feeding, watering, trapping and holding animals. Each type had to be designed to adapt to the natural island terrain and to minimize the extent of environmental impact. Each has proven itself functional for the purpose for which it was designed, and each continues to be successfully utilized in our operations. The development of capture systems, which is still evolving, is becoming more sophisticated and, as it does, our ability to trap and monitor the animals constantly improves.

The expansion of the laboratory facilities on the mainland (Summerland Key), and the development of a holding facility with individual caging on the island, will greatly improve our ability to monitor individual animals. Hopefully, over a period of time, this will include every animal on the island. All the animals trapped for shipment will be tuberculin tested. Direct rectal cultures and fecal enrichments will be tested for the presence of enteric pathogens. A blood sample will be drawn and serum sent to the Wilmington laboratory and tested for antibodies to herpes B-virus.

As previously stated, we have established, through our current monitoring schedule, that the animals remain free of antibodies to herpes B-virus, tuberculosis, salmonella and shigella. The potential for microbial contamination by birds and animals is recognized but is nevertheless minimal because there is no indigenous animal population to serve as a reservoir of infection.

Daily observations of animals and the use of a master file have enabled us to document animal inventory, reproductive state (estrous, pregnant, infant), physical condition and location on the island at time of observation. Troop organization, movements and territorial establishments are being documented. This information takes on greater significance now that the final group of animals has been introduced to the island in April 1976. This means that some reorganization of groups will most likely occur during the next breeding season resulting in a more normal social organization of the entire island population.

Reproduction continues to exceed our projections. The estimated time frame was based on the age and time of introduction of females to the island and the time for maturation of the males. It is our belief that the small losses from stillbirth or abortion are at least partially due to the elimination of potential pathogens through the rigorous quarantine, the lack of natural predators and, most of all, the high nutritional status of the animals and the lack of competition for food.

It is estimated that the island will be at 80% of anticipated production through the 1977 breeding season and at an annual production of 800-900 infants by 1978 or 1979.

The program monitoring colony health status is being significantly expanded because of the completion of the island holding area, complete laboratory capability plus the addition of a full-time laboratory technician. Also, during the trapping and holding of animals for the health testing program, the infants that are with their mothers will be permanently identified. The infant's number and the female's number will be recorded so that the maternal parentage can be identified on future breeders.

Along with the above, and throughout the continuous observation and documentation of a female's reproductive status, we can begin to cull and replace non-breeding females with animals that were born on Key Lois. Over a period of time, the entire breeding colony will then consist of animals produced on Key Lois. We feel that, because of the excellent production we have experienced, the replacement program will begin sooner than expected. Food and water consumption has been standardized. The average consumption per monkey is 0.23 kg of food and 500 to 800 cc of water depending on the time of the year and the weather. Supplying the island has been improved through the use of two shallow draft boats fitted with water carrying tanks, a large walk-in cooler at the base house for food storage, and continuous food-ordering procedures.

Daily observations supply information for the inventory and for some behavioral characteristics. The periodic introduction of new groups of animals, primarily juvenile and young adult females, has caused almost continual reorganization of the island population. However, the last group of animals was introduced to the island in April 1976. These animals are now being conditioned to the island environment. During their release and the ensuing breeding season, we should observe the stabilization of troops and obtain more complete behavioral data.

Three methods are currently being used to measure environmental impact on and around the island. In addition to visual observation by the people involved, forty fixed points on the island are photographed monthly using 35 mm color photography. This permanent record is used to compare identical locations against previous photographs. Also, on a monthly basis, since there was concern about surrounding water contamination, water samples are collected and tested by our laboratory and by an independent laboratory for the presence of fecal coliforms. Other than expected mechanical damage to the trees by the monkeys, no adverse effects to the environment have been noted.

To date, the colony on Key Lois, Florida, comprises in excess of 1,800 animals. This includes approximately 1,350 breeders and over 500 infants suckling to two years of age. Minimal fetal and infant losses have been experienced. In addition, some 50 to 100 animals are currently being harvested and will be supplied to the research community in the fall of 1976.

There are many advantages of island free-ranging breeding of rhesus monkeys. One of the most important is that in conventional types of breeding facilities, either totally indoor or partially indoor, the cost of the monkeys increases each year because of labor intensification, energy cost increases, and replacement of mechanical equipment. It is estimated by NIH that the ratio of people to monkeys is one person per 75 to 125 monkeys, depending on the program. We currently have three people taking care of approximately 1,800 monkeys. One is principally responsible for facilities maintenance, leaving two people to feed, water, capture, collect samples for quality control and treat animals if necessary.

It is entirely conceivable that a second island of similar or increased size could be serviced with current equipment and from the same base of operation. The only necessary addition would be two or possibly three people. This would mean that as the number of monkeys to be maintained and bred increases, the per-monkey cost of offspring could conceivably decline or at the very least remain constant. We feel that this can eventually be realized even with the improved quality of animal being supplied soley from Key Lois. In another 4-5 years, many of these forecasts will either be confirmed or found inaccurate.

In addition, the exercise, natural habitat and lack of competition for food and water in a free-ranging breeding situation produced monkeys of greater vigor and size for their age. It reduces the possibility of the transmission of airborne infections because of the open environment as compared to a more confining situation, and produces a model which behaviorally would be most similar to that found in nature.

It is true that a monkey cannot be located instantly, even though every animal is tattooed; but, as a practical matter in the three years of operation, there has been no specific situation where a monkey of a given sex with a specific number has had to be captured at a moment's notice.

The potential problems of contamination from extraneous sources and, of course, of a major tropical storm are present. However, to date we have fortunately not been subjected to these hazards; and we are proceeding according to our established plan.

ACKNOWLEDGEMENTS

The author wishes to express his appreciation to Mr. Barry R. Sherman for his assistance with the observations and collection of the information on behavior and reproduction. Study partially supported by Division of Research Resources, Animal Resources Branch, National Institutes of Health, Contract No. N01-RR-6-2137.

REFERENCES

Foster, H.L. (1975). *In* "Breeding Simians for Developmental Biology" (F.T. Perkins and P.N. O'Donoghue, eds), pp. 107-117. Laboratory Animals, London.
Foster, H.L. (1976). *Lab. An. Sci.*, 26, 374-382.
Orcutt, R.P., Pucak, G.J., Foster, H.L., Kilcourse, J.T. and Ferrell, T. (1976). *Lab. An. Sci.*, 26, 70-74.
Pucak, G.J., Orcutt, R.P., Judge, R.J., Foster, H.L. and Rendon, F. (1974). Presented at 25th Annual Session American Association for Laboratory Animal Science, October 21st. (Abstract).
Pucak, G.J. and Foster, H.L. (1975). *In* "Proceedings of the National Cancer Institute Symposium on Biohazards and Zoonotic Problems of Primate Procurement, Quarantine and Research". *Cancer Research Monograph Series*, (M.L. Simmons, ed.), pp. 109-127.

A PROGRAMME OF PREPARTUM CARE FOR THE RHESUS MONKEY, *MACACA MULATTA*: RESULTS OF THE FIRST TWO YEARS OF STUDY

C.J. MAHONEY and S. EISELE

Wisconsin Regional Primate Research Center, University of Wisconsin, Madison, Wisconsin USA.

An incidence of foetal death of 12 to 19% has been reported in laboratory housed *M. mulatta* (Valerio et al., 1968; Wagenen, 1972). In our own non-experimental breeding colony of rhesus monkeys, most of which are singly caged, the rate of foetal loss ranged from 15 to 21% annually over three consecutive years (1972-74).

No systematic study has been made to identify the causes of foetal death in large colonies of non-human primates. Such investigation is important not only clinically but in experimental studies of foetal development and pregnancy. To understand these problems better, a continuing programme of prepartum care was begun two years ago to establish criteria of maternal and foetal well-being during late pregnancy and to determine the causes of foetal morbidity and mortality.

Beginning on day 150 of gestation, or earlier if an obstetrical problem was suspected, manually restrained, unanaesthetized females were examined at 1 to 5 day intervals. Examination procedures included (a) bi-manual rectal examination to monitor palpable changes in the *cervix uteri*, and to determine foetal presentation, engagement and movements, (b) vaginoscopic observation of the cervix, and (c) ultrasounding by a doppler blood flow monitor (Model BF4A, Medsonics Incorporated, Mountain View, California 94042 USA), to detect foetal heart sounds, the technique of which has been described (Mahoney and Eisele, 1976).

Imminence of parturition was most reliably determined by monitoring changes in physical characteristics of the *cervix uteri*. As in *M. fascicularis* (Mahoney, 1975), the cervix of *M. mulatta* passes through 10 distinct phases of textural change during pregnancy, as determined by rectal palpation. From a firm structure, in early gestation, the cervix progressively softens until, by the last 1 to 3 days of pregnancy, it becomes completely non-palpable. The rate at which these changes occurred varied among animals. However, statistical analysis indicated a negative linear regression ($r = 0.73$, $p < 0.01$) between the cervical phase and the number of days preceding parturition (Mahoney and Eisele, in preparation).

Vaginoscopic examination of the cervix indicated that complete effacement of the *portio vaginalis* and dilatation of the external os (> 1.5 cm) occurred within 1 to 4 days before parturition. Dilatation of the internal os did not occur until 12 to 24 hours before delivery. In this event, the opportunity was readily afforded to examine the chorioamniotic membranes for meconium staining, a sign of foetal distress (Hon, 1969). An increased secretion of an oestrogenic type of cervical mucus occurred during the last 10 days of

TABLE I

Pregnancy rate and incidence of foetal viability and death in a non-experimental breeding colony of M. mulatta

Type of Delivery		Pre-Study Year	Study Year I	Study Year II
Viable	Vaginal	84 (74.3)	108 (79.4)	101 (85.6)
	Caesarean	5 (4.4)	7 (5.2)	7 (5.9)
	Total	89 (78.8)	115 (84.6)	108 (91.5)
Dead	Vaginal	15 (13.3)	9 (6.6)	7 (5.9)
	Caesarean	3 (2.7)	3 (2.2)	2 (1.7)
	Abortion	6 (5.3)	9 (6.6)	1 (0.9)
	Total	24 (21.3)	21 (15.4)	10 (8.5)
Total Number of Pregnancies		113	136	118

Figures in brackets indicate percentage of total pregnancies per year.

pregnancy, occasionally being blood tinged in the last 12 to 24 hours.

Diagnosing foetal viability by detecting spontaneous or elicited body movements was found to be unreliable. In contrast, ultrasonic monitoring to detect foetal heart beats proved to be a reliable and simple technique (Mahoney and Eisele, 1976). Detection of foetal heart beats within 2 days or less of stillbirths indicated that the majority of late term foetal deaths occurred in the immediate preparturient period.

Caudal presentation of the foetus was regarded as potentially dangerous if the cervix had reached an advanced stage of softening, indicating impending parturition. Because of the risk of breech delivery, rotation of the foetus into the cephalic presentation was accomplished by bi-manual rectal palpation. In some instances, several daily repeated manoeuverings were required. The foetal heart rate was counted before and after each manipulation to determine whether strangulation of the umbilical cord had occurred.

Pelvic engagement of the foetus occurred 1 to 3 days before delivery. This, together with complete softening and dilatation of the cervix, was a reliable index of approaching parturition.

In the pre-study year 113 pregnancies occurred, compared to 254 in the 2 years of study. The incidence of viable vaginal deliveries increased from 74.3 to 85.6%.

A total of 14 live infants was delivered by hysterotomy during the study period. Knowledge of hindsight suggested that 5 deliveries were clinically unwarranted. However, the remaining deliveries were deemed necessary because of *placentae abruptio* and *placenta praevia* (1 case each), pathological contraction ring (1 case) (Reid, 1972), uncorrectable malpresentation of the descended member of twins (1 case), severe amniotic meconium staining indicating *in utero* distress (4 cases), and severe maternal vaginal prolapse (1 case). One of the cases of *in utero* foetal distress was diagnosed by an oxytocin challenge test (Freeman, 1975).

Twenty-one late-term foetal deaths (> 139 days gestation) occurred over the 2 year study period, 5 of which were delivered by Caesarean section. The causes of death, and the number of cases, indicated in parentheses, were as follows: breech (7) and *fruntum dystocia* (1), *placenta praevia* and *placentae abruptio* (1 each) and intramural placental infarction or haemorrhage (5). The causes of the remaining 6 deaths were undetermined. Prompt surgical intervention would have avoided 5 stillbirths. Foetal hypoxia, due to placental insufficiency (Hon, 1969), was indicated in 3 pregnancies by late deceleration of the foetal heart rate relative to the onset of uterine contractions induced by the intravenous infusion of pitocin (6 units/hour). In the remaining 2 pregnancies, a non-haemorrhagic *placenta praevia* and a breech presentation were diagnosed several hours before stillbirth.

Haematological and bacteriological examinations indicated that approximately half of the early gestational foetal deaths (< 140 days) were associated with maternal infections, principally shigellosis. Abortions decreased in the last 18 months of study concomitant with a general decline in incidence of shigellosis throughout the colony.

Possibly, the high incidence of stillbirths due to foetal malpresentation is associated with the lack of exercise of individually caged females. Further study is necessary to determine whether foetal death is related to maternal parity.

ACKNOWLEDGEMENTS

The authors are grateful for the technical help of Michael Hempel and Mark Capriolo (Wisconsin Primate Laboratory) and to Ms. Stephanie Shepard for data and manuscript preparation.

REFERENCES

Freeman, R.K. (1975). *Amer. J. of Obstet. Gynec.*, 121, 481-489.
Hon, E.H. (1969). *In* "Controversy in Obstetrics and Gynecology" (D.E. Reid and T.C. Barton, eds), Chap. 12. W.B. Saunders and Co., Philadelphia.
Mahoney, C.J. (1975). *Lab. Anim. Handb.*, 6, 261-274.
Mahoney, C.J. and Eisele, S. (1976). *J. Med. Primat.*, 5, 284-295.
Reid, D.E. (1972). *In* "Principles and Management of Human Reproduction" (D.E. Reid, K.J. Ryan and K. Benirschke, eds), pp. 552-585. W.B. Saunders and Co., Philadelphia.
Valerio, D.A., Courtney, K.D., Miller, R.L. and Pallotta, A.J. (1968). *Lab. Anim. Care*, 18, 589-595.
Wagenen, G. van, (1972). *J. Med. Primat.*, 1, 3-38.

BREEDING STATISTICS OF *MACACA FASCICULARIS* IN BASEL ZOO

W. ANGST

Zoological Institute, Basel University, Switzerland.

INTRODUCTION

In the course of a longitudinal behavioural study of the crab-eater colony in Basel Zoo, reproductive data have been collected from 1961 to the present. Due to the author's absence during 1968, 1969 and 1970, no data are available for this period. The data presented here cover 12 years, 1961-1967 and 1971-1975. The group size varies from 58 to 99 individuals. No additional monkeys were introduced after 1949. In fact, high reproductive output allowed the sale of 106 individuals from 1st January, 1962 until 31st December, 1975 (for details of maintenance and group composition, see Angst, 1974 and 1975).

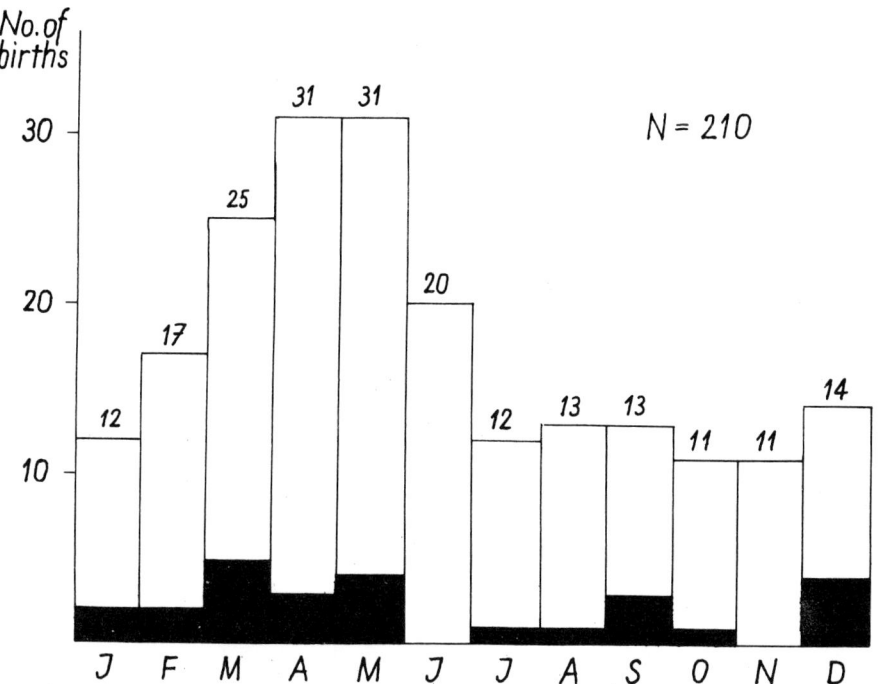

Figure 1. *Monthly distribution of births (years 1961-1967, 1971-1975). Black, still births; White, live births; Numbers, totals.*

TABLE I

Sex distribution of births

	Male	Female	Sex not determined
Live born	90	83	11
Stillborn	10	9	7
Total	100	92	18

TABLE II

Age at first delivery (11 females)

Minimum:	3 years 4½ months
Mean:	4 years 4 months
Maximum:	5 years 9 months

TABLE III

Eating of placenta after first and consecutive deliveries (75 cases, 1973-1975)

	Placenta eaten	Placenta not eaten	Not classifiable
First delivery Live births	5	5	1
Not first delivery Live births	47	10	4
Stillbirths	2	1	-

TABLE IV

Survival ratios for the first year of life (infant killing excluded)

	Survivors	Deaths	Ratio
Males	59	14	4.2
Females	45	15	3.0

TABLE V

Seasonal survival ratios for the first year of life (infant killing excluded)

	Survivors	Deaths	Ratio
Winter (Nov-Feb)	26	12	2.2
Rest of year (Mar-Oct)	78	27	2.9

TABLE VI
Birth intervals in days

	Median	Mean	S. dev.	Range	N
After stillbirth	376	381.3	114.5	186-538	16
After loss of infant during its first month of life	363	404.8	143.5	231-696	23
After loss of infant during its 2nd-6th month of life — all values	346	434.7	295.3	218-1201	9
— with exclusion of the maximum value	342	338.9	72.6	218-469	(8)
After loss of infant during its 7th-12th month of life	432	437.0		408-471	3
After raising infant until next birth or at least for one year	436	484.6	150.8	299-982	70
				Total	121

RESULTS

Births occur throughout the year but, as Figure 1 shows, there is a marked peak in March, April and May. Out of 210 births there were 26 stillbirths, or 12.4% of all births. There were almost as many females as males born, for both live and stillborn infants (Table I). Table II summarizes the age of 11 females at their first delivery including two 1976 births. The case of the female who gave birth when only 3 years 4½ months old implies a conception at the age of 2 years 11 months. Multiparous females eat the placenta significantly more often than not, but in half of the cases primiparous females did not eat the placenta (Table III). The difference in live births between first deliveries and later ones approaches statistical significance (X^2= 3,459 df = 1, p<0.05). If selective killing of female infants is excluded females still have a lower survival rate (though not statistically significantly) than males (Table IV). During winter the monkeys can enter a heated house. Some animals often remain outside, as it is somewhat crowded inside. This may account for the slight, though not statistically significant, lower survival rate for winter births as compared to births at other times of the year (Table V). Birth intervals are shorter after still births or early infant loss than after successful rearing of infants (Table VI). The difference between birth intervals for still born or lost infants within the first half year (N = 48) and birth intervals where the infant survived for more than half a year (N = 73) is significant at the 0.01 level (t-test).

Do high ranking females have higher reproductive success than low ranking ones? There are at least two years of data after the year of the first delivery for 30 females. Due to changes in the dominance hierarchy during the observation period, the 12 highest ranking and the 12 lowest ranking females are compared. This distinction eliminates the possibility of including dominance reversals.

In 66 female-years the dominant females have produced 43 live births, a yearly reproduction rate of 65.2%. The subordinate females in 69 female-years have produced 46 live births, which gives a yearly rate of 66.7%. The yearly reproduction rate of all 30 multi-parous females is 65.6%. So it can be concluded that under the given conditions of captivity females of different rank reproduce equally well.

ACKNOWLEDGEMENTS

I wish to express my gratitude to the staff of Basel Zoo for their kind co-operation: the Director, Professor E.M. Lang; the Vice-Director, Dr. H. Wackernagel; the Veterinarian, Dr. D. Rüedi; the Keepers, G. Ruby and M. Giuliani. For his valuable help during recent years I especially thank D. Thommen. Finally I thank J.A. Kurland for help with the English version of this paper.

REFERENCES

Angst, W. (1974). *Adv. Ethol.*, <u>15</u>.
Angst, W. (1975). *In* "Primate Behavior" (L.A. Rosenblum, ed.), Vol. 4, pp. 325-388. Academic Press, New York.

THE INFLUENCE OF CHANGED PHOTIC CONDITIONS ON THE BREEDING PERFORMANCE OF LABORATORY PRIMATES (*MACACA MULATTA*)

FREDERICK E. BIRKNER

*New York University, School of Medicine,
400 East 34th Street, New York, NY 10016 USA.*

To meet investigators' needs for *neonate* rhesus monkeys (*Macaca mulatta*) of known gestational age, the Institute of Rehabilitation Medicine of New York University established a breeding colony on Welfare Island (now Roosevelt Island), an island in the East River at New York City. Forty-two wild-born female M. *mulatta* weighing from 4.5-5.0 kg, procured from a commercial source, provided the foundation stock. Animals were judged to be mature and of breeding age by criteria of weight and dentition. They were maintained under conventional husbandry conditions in single cages conforming to governmental standards. Many animals were anoestrous for a period from 6-12 months. After an 8 month stabilization period, approximately 80% of the animals showed menstrual cycles varying in length from 26-30 days. They were taught to present the perineum on signal for observation. Animals showing persistent amenorrhoea were culled from the program and replaced with other mature females. Protocol of the breeding program was largely based on the findings of Van Wagenen who reported that days 10-13 after the onset of menstrual flow constitute the most fertile period of the M. *mulatta* menstrual cycle, the modal point being determined by the length of the regularly observed cycle of individual animals. Accordingly, those showing a menstrual cycle of 28 days were introduced to a proven compatible male on day 11 after menstrual flow had been observed. Those with regular cycles of 26 days were mated on day 9; when a 27-day menstrual cycle was noted the day of mating was 10. Animals with cycles of 20 or 30 days were mated on day 12 and 13, respectively. Mating took place in the males' home cage, which was placed in a breeding room separate from the rest of the colony. The pairs were left together for 48 hours. Presence of a seminal plug was assumed to be a confirmation that copulation had taken place. Conceptions were confirmed approximately 5 weeks later by rectal palpation of the uterus. Infants delivered, by caesarean section or vaginally, were removed and raised under conditions of intensive care. No remating was attempted during post-partum oestrus but females were mated, usually with the same sire, during the following menstrual cycle(s). By the spring of 1969 the breeding colony consisted of 54 females and three proven males.

During the first three years of the period reported here, the breeding colony had the benefit of exposure to daylight through windows which covered the entire west wall of the secondary enclosure. From 0830 to 1630 hours illumination was supplemented by incandescent lighting. Average footcandle (fc) intensity recorded at the frontal plane of the cage level ranged from a low of 9 fc at 1600 hours during the month of December to a high of 20 fc at noon

TABLE I

Conception data.
Conceptions during exposure to daylight

	May	Jun	Jul	Aug	Sept	Oct	Nov	Dec	Jan	Feb	Mar	Apr	Total
Daylight hours (h) and minutes (m) at 40°N Latitude	445.46	448.54	455.36	425.43	374.18	346.16	304.13	291.31	300.12	298.50	369.45	397.42	
1969-1970	4 (11.4)	5 (14.3)	3 (8.6)	2 (5.7)	1 (2.9)	4 (11.4)	3 (8.6)	4 (11.4)	2 (5.7)	2 (5.7)	3 (8.6)	2 (5.7)	35 (100.0)
1970-1971	5 (15.6)	2 (6.2)	5 (15.6)	4 (12.6)	2 (6.2)	2 (6.2)	1 (3.1)	1 (3.1)	1 (3.1)	3 (9.4)	3 (9.4)	3 (9.4)	32 (100.0)
1971-1972	3 (10.7)	2 (7.1)	4 (14.3)	1 (3.6)	2 (7.1)	4 (14.3)	2 (7.1)	4 (14.3)	2 (7.1)	0 (0.0)	2 (7.1)	2 (7.1)	28 (100.0)
Total	12 (12.6)	9 (9.5)	12 (12.6)	7 (7.4)	5 (5.3)	10 (10.5)	6 (6.3)	9 (9.5)	5 (5.3)	5 (5.3)	8 (8.4)	7 (7.4)	95 (100.0)
Rank daylight hours	3	2	1	4	6	8	9	12	10	11	7	5	
Rank conceptions	1	3	1	5	7	2	6	3	7	7	4	5	

Figures in brackets represent percentage.
May-October. 56% of all daylight hours - 57.9% of all conceptions
November-April. 44% of all daylight hours - 42.1% of all conceptions
$\chi^2 = 4.85$, df = 1, $p > .05$.

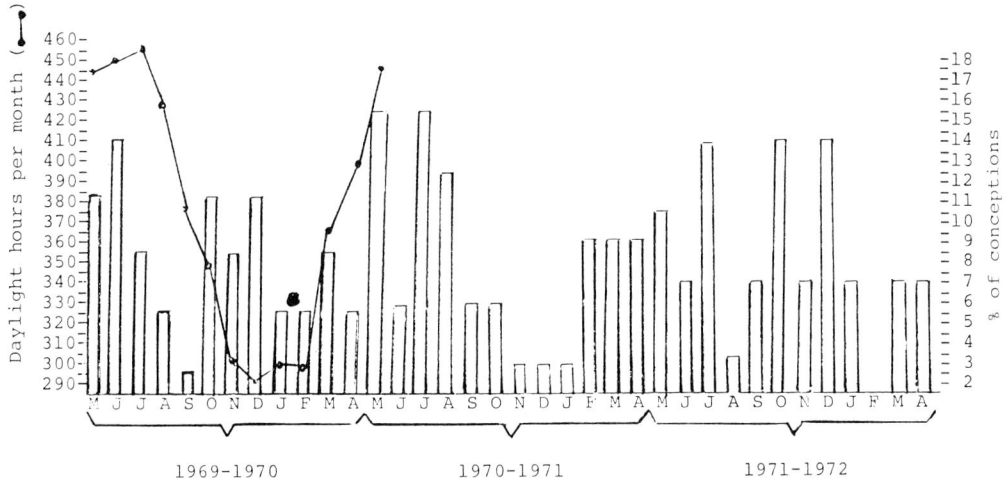

Figure 1. *Relationship of daylight hours to conceptions.*

during May and July.

Length of day at 40°N latitude (New York City) varies little from year to year during corresponding calendar months. The fluctuations do not appear to be significant for the purpose of this report. At 40°N latitude sunrise never occurs later than 0830 hours and sunset never earlier than 1630 hours and while supplemental lighting may have increased the intensity of light in the animal room it did not vary the length of day indigenous to the region.

At the end of April 1972 the colony was moved to new quarters at the New York University Medical Center in downtown Manhattan. The animal rooms were without windows and were climate controlled. Ambient temperature and humidity levels were stabilized (as in the animals' previous quarters) at 25.5° (± 3) and 45-55% RH respectively. Lighting was fluorescent, measuring 12 fc at cage level. A 14/10 hour light/dark cycle was maintained. Animals did not appear to be stressed by the move. Most of them had made the trip to the downtown location on previous occasions for laboratory tests and other procedures. They were handled by their accustomed caretakers and housed in their familiar cages. Only three of the animals failed to cycle regularly during the first few months at their new quarters. By July 1972 all animals had returned to their normal menstrual cycle.

During the period when the monkeys were exposed to daylight, 59.9% of all conceptions took place during the period May-October, the portion of the year which has 56.0% of all annual daylight hours. Difference in the conception rate for this period and the months November through April is statistically significant to the .05 level (χ^2).

For the two years under artificial light and the 14/10 hour light/dark cycle the pattern of conceptions remained relatively constant with 56.9% of conceptions recorded during the May-October period.

Southwick et al. (1965) reported that in free-ranging *M. mulatta* in northern India, most copulations were observed during October and November. Vandenberg and Vessey (1968) indicated that on Cayo Santiago (Puerto Rico) matings begin in August or September and reach their peak during November and December. Other investigators (e.g. Johannson et al., 1968) observed a summer anovulatory period in laboratory-housed *M. mulatta*. Reviewing the cumulative conception record of our colony we found conceptions occurring throughout the year

TABLE II
Conception data.
Conceptions during exposure to fluorescent light only (14/10 hr light/dark cycle)

	May	Jun	Jul	Aug	Sep	Oct	Nov	Dec	Jan	Feb	Mar	Apr	Total
1972-1973	2 (6.9)	2 (6.9)	4 (13.8)	0 (0.0)	4 (13.8)	5 (17.2)	2 (6.9)	3 (10.3)	1 (3.4)	0 (0.0)	2 (6.9)	4 (13.8)	29 (100.0)
1973-1974	0 (0.0)	1 (4.3)	3 (13.0)	2 (8.7)	1 (4.3)	4 (17.4)	2 (8.7)	3 (13.0)	3 (13.0)	2 (8.7)	0 (0.0)	2 (8.7)	23 (100.0)
Total	2 (3.8)	3 (5.8)	7 (13.5)	2 (3.8)	5 (9.6)	9 (17.3)	4 (7.7)	6 (11.5)	4 (7.7)	2 (3.8)	2 (3.8)	6 (11.5)	52 (100.0)
Rank conceptions	7	6	2	7	4	1	5	3	5	7	7	3	

Figures in brackets represent percentages.
May-October. 53.8% of conceptions.
November-April. 46.2% of conceptions.

Macaca mulatta

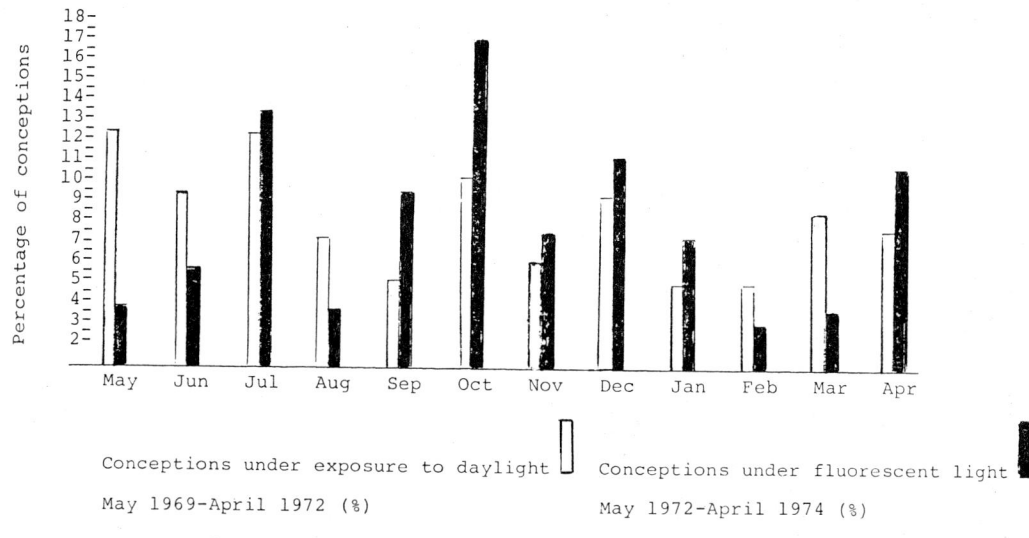

$\chi^2 = 2.42$, df=1, $p < .12$ (not statistically signigificant)

Figure 2. Summary of conceptions.

with a seasonal peak during the May-October period. No statistically significant difference in conception rates was noted between the two different photic conditions under which this breeding program was conducted. It does appear then that diurnal variations alone were not responsible for the paradigm of conceptions noted.

REFERENCES

Johansson, E., Neil, J.D. and Knobil, E. (1968). *Endocrinol.*, 82, 143-148.
Southwick, C.H., Beg, C.H. and Sidiqui, M.R. (1965). *In* "Primate Behavior" (I. Vore, ed.). Rinehart and Winston, New York.
Vandenberg, J.G. and Vessey, S. (1968). *J. Reprod. Fert.*, 15, 71-79.

BREEDING OF MACAQUES AND CHIMPANZEES AT THE DUTCH PRIMATE CENTRE

W. VAN VREESWIJK AND H. KONING

The Primate Centre TNO, 151 Lange Kleiweg, Rijswijk ZH, The Netherlands.

One of the most important research projects of the Dutch Primate Centre is the study of the immunogenetics of tissue antigens of rhesus monkeys and chimpanzees. These species are used as optimal models for pre-clinical research, primarily in transplantation biology. For the immunogenetic work, large families are needed and therefore breeding colonies of both species had to be established at the Centre*.

The breeding programme for rhesus monkeys (*Macaca mulatta*) was started in 1965, for chimpanzees (*Pan troglodytes*) around 1970. The principal method of breeding macaques, the so-called rotating gang breeding, has been described previously (Goosen, 1972b; Balner, 1975). Since the supply of rhesus monkeys from the countries of origin has been seriously curtailed in the last few years, the development of more economical and, if possible, more efficient breeding systems has become an important issue.

In this report, information will be provided about the various systems currently used for the breeding of rhesus monkeys at the Rijswijk Centre. Attention will be paid also to the assays used for early pregnancy detection by *in vitro* and *in vivo* methods. Furthermore, a brief account will be given of the raising of chimpanzees at the Centre. It will be shown that the reduction of infant mortality during the first year of life, by the use of an isolation unit in the nursery, has led to a very satisfactory production record for this relatively small but very valuable breeding colony.

BREEDING OF RHESUS MONKEYS

The initial interest in the breeding of rhesus monkeys was the production of large families to obtain maximal numbers of related animals of known age and genetic background for various immunogenetic research programmes. In the early period (about 1965-1972) only a conventional harem-type gang breeding system was used and results obtained with that method have been published (Goosen, 1972b). Lately, for reasons already mentioned (scarcity), the application of more efficient breeding systems became imperative. The advantages and disadvantages of the various methods which are currently in use will be briefly discussed in the following sections.

---

*The Centre also maintains smaller breeding colonies of stump-tailed macaques (M. arctoides) and marmosets (C. jacchus) but those will not be considered in the current communication.

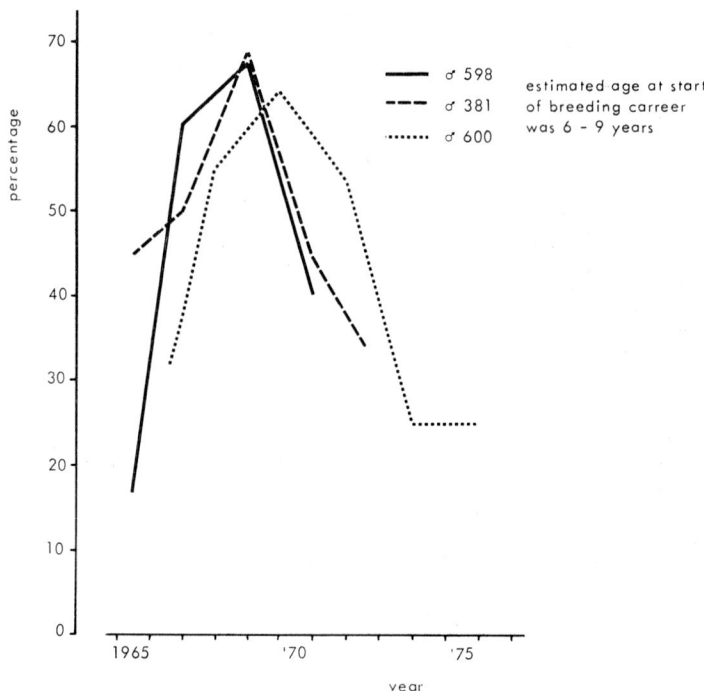

Productivity is expressed as the percentage of pregnant females in each individual harem; indicated values are pooled data for every 2 years (see further text).

Figure 1. *Productivity of 3 male rhesus monkeys used in a rotating gang breeding system (average of 20 females per harem).*

Breeding with imported rhesus monkeys

Rotating gang breeding. This type of breeding has been routine at the Centre for many years and has been reported in detail (Goosen, 1972b; Balner, 1975). Briefly, each male has a harem of about 20 females which are assigned to him exclusively. Four or five females are exposed to the male during mating periods of 3 weeks. Six weeks after separation of this first group and introduction of a second group, possible pregnancies in the first cohort of animals are diagnosed by rectal palpation. Pregnant animals are kept apart and the others are returned to the harem for recycling. Females which do not become pregnant within two years are assigned to another male for a year. If this change does not lead to pregnancy, the female is given one more chance with a third partner before being permanently removed from the breeding colony. Weak or diseased breeders are, of course, also removed from the harem.

Results obtained with this type of breeding, using 17 males and about 200 females, showed an average conception rate (i.e. the number of pregnant females divided by the total number of females actively participating, x 100) of 55%, with about 10% abortions or still-births. This conception rate may seem low. However, as already indicated, the Centre is primarily interested in raising families with a maximal number of full sibs; this implies that breeders will be kept in the colony until a very advanced age which can drastically reduce the overall conception rate (see also below).

To get a better insight into factors possibly influencing productivity of the males, their yearly productivity was analysed. The yearly conception rate of the females of each harem (in percentages, as defined above) was used as criterion. Results obtained for 3 males are graphically presented in Figure 1. It is apparent that in this type of breeding system there is a peak in the productivity for each male. This peak is estimated to occur when the animal is 10-15 years of age, the highest score was found to be 83%. Furthermore, the analysis revealed that, unlike what has been reported by Valerio et al. (1971), there is a distinct seasonal influence on the productivity of the females (not shown in the figure). The highest fertility clearly lies between September and June and this holds true also for animals which have been kept in the laboratory for more than 5 years.

Long-term gang breeding. This is a slightly modified version of the described rotating system. Here, one male and a group of 8-10 females are introduced into the breeding pen at the start of the season (September 1st). Two months later and subsequently every month, the females are examined for pregnancy by rectal palpation. Pregnant animals are separated and replaced by non-pregnant members of the same harem (a harem again usually consists of about 20 animals). With this breeding method the proportion of injured females is significantly lower than when the rotating system is used. An explanation might be that the social rank order between the animals is more rapidly established and remains more stable in the former type of breeding system, while four or five 'new' females are periodically introduced in the rotating system. According to the limited experience at the Rijswijk Centre, the conception rate reached with this type of long-term gang breeding was, if anything, higher than when the rotating system of gang breeding was applied (averages of about 70% and 50%, respectively).

Timed matings. In a special breeding programme* in which animals were immunized with a variety of placental extracts (to test their effect on ferility), timed matings were required. In these experiments, 54 of 63 females were immunized and a total of twenty-six became pregnant. This yield may again seem rather low. However, when the number of pregnancies was related to the number of exposures (mating periods), the success rate was about 20%, which is similar to what has been reported by others using timed matings (Valerio et al., 1971). Furthermore, a possible influence of the immunizations on the recorded pregnancy rate cannot be excluded.

In order to record the menstrual cycles, daily vaginal swabs were taken. The majority of the females had a cycle of 20-26 days. Ten or 11 days after the onset of menstrual bleeding, the females were mated for 3 or 4 days. Usually a single male was caged with a single female; occasionally the male had to be caged with 2 or 3 females simultaneously (lack of sufficient male breeders). After such mating periods, females were removed from the cage and kept separately until 10 or 11 days after the onset of the next cycle; at that time they were again exposed to the male (always the same partner). Data obtained so far (observations in 1976, only) suggest that the conception rate

Performed at the Rijswijk Centre (in a collaborative project) for the Task force on immunological methods of fertility regulation of WHO, Geneva.

TABLE I

Comparison of results obtained with an in vitro *method and by rectal palpation to diagnose pregnancies in rhesus monkeys*

Mating period	Animals (code number)	Day after termination of mating period on which urine sample was taken							Pregnancies according to rectal palpation	Proved pregnancies (deliveries or abortions)
		16	17	18	19	20	21	22		
three days	1				?					
	2				?		?			
	3				?		−			
	4				?		−			
	5				?		−			
	6				?	−				
	7				?					
	8				?					
	9			?			+		yes	yes
	10						+		yes	yes
	11			?			+		yes	yes
	12				+		+	+	yes	yes
	13			−		+		+	yes	yes
	14				+		+		yes	yes
	15						+		yes	yes
	16						+		yes	yes
	17			+		?			yes	yes
	18		−	+					yes	yes
	19			+					no	no
four days	20				+				yes	yes
	21	?			+				yes	yes
	22		+	+					yes	yes
	23			+					yes	yes
	24	?	+						yes	yes
	25	+			−				yes	yes
	26				?				yes	yes
	27	?	−						yes	yes
	28			?		−			yes	yes
	29		−						yes	yes

In 45 additional cases, all *in vitro* tests and results of rectal palpation were negative and the animals had indeed not become pregnant.

with this timed mating system is higher than the conception rate attainable with the rotating gang breeding system which has been routinely applied for the past decade. A meaningful comparison with the conception rate obtained with the long-term gang breeding (also described above) is not yet possible.

The application of the timed-mating system made it possible to test the predictive value of an *in vitro* pregnancy test for non-human primates, described in 1974 (Hodgen et al.). About 3 weeks (18-23 days) after a mating period, fresh urine samples are taken and the *in vitro* test performed. Six weeks after the mating period, the females are examined for pregnancy by the

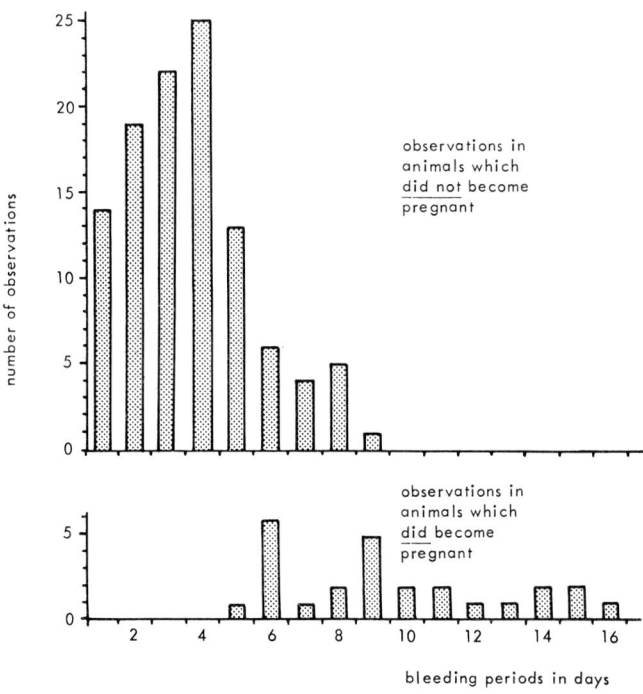

Each shaded rectancle represents an observation. The total number of pregnant females studied was 26. See further text.

Figure 2. Variation in the duration of the first bleeding period in female rhesus monkeys after timed-mating episodes of 3 or 4 days.

usual rectal palpation method*. Table I represents a comparison of results obtained with the *in vitro* and *in vivo* method of diagnosing pregnancies in rhesus monkeys. As can be seen, there was a high positive correlation between results obtained with both types of examination, especially when urine samples were taken 21 days after the mating period. This implies that with the *in vitro* assay, pregnancies can be reliably diagnosed 3 weeks earlier than when the conventional palpation method is employed.

Furthermore, a phenomenon was observed which had been described earlier by Valerio et al. (1969), namely that bleeding periods observed after matings in which females had actually become pregnant were significantly prolonged. This is shown by the histograms of Figure 2. Interestingly, in the majority of cases in which pregnancies occurred (20 of 26), the bleeding period was fragmented, i.e. there was an interval of 1-3 days in which bleeding was

If performed by experienced individuals, this type of examination has a very high predictive value: at the Rijswijk Centre, only 14 of 732 examinations were 'false' positives or negatives in the course of the past 2 years.

TABLE II

Breeding performance of imported and laboratory-born male and female rhesus monkeys

	Imported male breeders Pregnancies (P) scored per year						Laboratory-born male breeders Pregnancies (P) scored per year					
	1435			2315			EE		V			
	'74	'75	'76	'74	'75	'76	'75	'76	'73	'74	'75	'76
Fifteen imported female breeders	P	P	P	P	P	P	P	P				
	P	P	P	P	P	P	P	P				
	P	P	P	P	P	P	P	P				
	P	P	P	P	P	P	P	P				
	P	P		P	P	P	P	P				
	P	P		P	P	P	P	P				
	P	P		P	P		P	P				
	P	P		P	*	*	P					
	P	P		P		P	P					
	P	P		P		P	P					
	P		P	P		P	P					
	P				P							
	P			P								
		P			P							
Four laboratory-born female breeders							X	P	P	P	P	
							X	P	P	P		P
							X	P			P	*

Symbols P indicate pregnancies followed by deliveries; * stands for abortions; X means that a female was not included in the breeding programme in a particular year.
The complete 'harems' of the imported males 1435 and 2315 consisted of 21 females; six of harem 1435 and seven of harem 2315 never became pregnant in the years 1974-1976. Rotating gang breeding was used.
The laboratory-born male EE had a harem of 14 imported females (three of which did not become pregnant in 1975 and 1976; EE also was mated to 4 laboratory-born females (one not pregnant in 1975 and 1976). Long-term gang breeding was used.
The laboratory-born male V had a harem of only 4 laboratory-born females which were all half-sibs of V.

interrupted. Incidentally, these so-called implantation bleedings usually occur at the time when a normal menstrual cycle would have been expected to start.

Inbreeding of rhesus monkeys

In the past few years numerous F_1 animals at the Centre have reached sexual maturity. This has permitted the initiation of a systematic programme of inbreeding (child x parent, inter-sib and inter half-sib matings). After initial difficulties, the results have become encouraging. So far 21 inbred offspring have been born: 4 from father x daughter matings, 16 from half-sib matings and 1 from a brother x sister mating. The youngest laboratory-born male was 6 years old when producing the first offspring, the youngest female about 5 years. The programme was started with 8 laboratory-bred males, 4 of which turned out to be active breeders. The excellent breeding records for two of them are shown in Table II. On the other hand, some of the laboratory-born males had to be removed from the inbreeding programme because of excessive aggressiveness.

A difficulty encountered when using laboratory-bred females for breeding, was that about 60% neglected their first child. In 7 of a total of 12 cases, babies had to be removed from the mothers and were successfully hand-reared. Interestingly, this difficulty was never observed with imported females even if they had been imported at an early age and had grown up under the same conditions in the laboratory. When laboratory-bred mothers gave birth to their second or subsequent children, the neglect of offspring was not observed except in a single case. Table II demonstrates that laboratory-born males and females can be excellent breeders, although the experience gained so far is obviously still limited. It may be of interest to note that the productivity of the laboratory-bred male EE (now 10 years of age) increased from 47% in 1975 to 72% in 1976. If this trend continues for a year or two and his productivity then declines in the following years, our speculation on the most productive period in a male's breeding life (based on the productivity curve for imported males of unknown age, see Figure 1) may be supported by data from a male breeder of known age: for the time being, the breeding record of male EE gives some support to the theory that peak productivity may indeed lie around the age of 10-15 years.

BREEDING OF CHIMPANZEES

Since the chimpanzee is phylogenetically close to man, it is a particularly valuable species for the study of the immunogenetics of tissue antigens in relation to those of man. This phylogenetic closeness was one of the main reasons why about ten years ago, a chimpanzee colony was established at the Rijswijk Centre. To facilitate immunogenetic studies and in view of the decreasing availability of chimpanzees, a modest breeding programme was initiated in 1970. The method chosen for breeding chimps was rather similar to that routinely used for rhesus monkeys, namely a rotating gang breeding system. Some results of breeding chimps at the Centre have been published (Goosen, 1972a; Balner, 1975). Pregnancies were and are still successfully monitored by a test which measures the chorionic gonadotrophin level in the urine (Boorman et al., 1974). Table III summarizes the updated results of breeding chimpanzees at the Rijswijk Centre between 1970 and mid-1976. It can be seen that the overall conception rate has increased from about 30% in the early period to approximately 50% in 1975-1976. The high rate of initial abortions may be attributable to the fact that the majority of the females were hardly sexually mature when starting to participate in the breeding programme. A detailed account of the methods practised to rear the infant chimps at the Centre will be given elsewhere (Beyersbergen et al., in preparation). Suffice it here to indicate that babies which are accepted by their mothers remain with her until weaning at 3 to 5 months of age. Incidentally, it has been shown recently that infants weaned at 3 months of age are in a better physical condition than those left with the mother for a longer period. In any case,

TABLE III

Records of chimpanzee breeding at the Primate Centre TNO (Period 1970-1976)

	'70	'71	'72	'73	'74	'75	'76*
Approximate number of mature females 'participating'		10			17		22
Live births	-	2	4	5	8	9	5
Abortions or still-births	1	1	4	3	3	2	2
Deaths occurring in 1st year of life	-	1	2	3	3	-	-

*At the time of this writing (July 1976), 6 animals were diagnosed to be pregnant.

babies which are neglected are instantly removed from their mothers and hand-reared in a nursery. The latter is located in a separate small building and equipped with a simple isolation unit. Only specialized personnel who take care of the babies have access to this building. It is reasonable to assume that the institution of this isolated nursery, towards the end of 1974, may have been the main reason for the drastic drop in the mortality rate of infant chimps during their first year of life (Table III).

CONCLUSIONS

During the past decade, breeding of rhesus monkeys at the Dutch Primate Centre has been practised successfully on a fairly large scale. Since the immunogenetic studies carried out at the Centre require the availability of large families, harem-type breeding methods were mostly applied. The rotating gang breeding system was used routinely; it is more economical than the timed-mating system, although the yield obtainable with the latter method is clearly higher. A third method, the so-called long-term gang breeding, was tried in the last few years. Results are encouraging so that this type of macaque breeding may become the method of choice in the near future. An analysis of the productivity of male breeders, covering a period of about 10 years, suggested that the peak of male reproductive capacity (at least if the rotating breeding system is used) may lie between the age of 10 and 15 years. If this is confirmed, breeding of rhesus monkeys would be most efficient if males be-between the age of 10 and 15 years were used and their productivity checked every year. Furthermore, an analysis of the Centre's breeding records revealed that an established *in vitro* method to diagnose pregnancies (about 3 weeks after conception) has a high predictive value in the hands of the Rijswijk investigators.

The breeding of chimpanzees at the Centre, although practised on a much smaller scale than for macaques, is also fairly successful. With a harem-type rotating breeding system, the overall conception rate is currently about 50% and thus similar to that observed in rhesus monkeys (if a similar rotating breeding system is used). A brief review of the short history of chimpanzee breeding at the Centre revealed that the initially high infant mortality was drastically reduced in the past two years. This coincided with (and may be attributable to) earlier weaning and, above all, to improved surveillance and health care in an isolated nursery.

ACKNOWLEDGEMENTS

The authors wish to express their sincere thanks to Dr. H. Balner for his invaluable suggestions during preparation of the manuscript, and Mr. F. Melk (at present a veterinary student at the Utrecht University) for conducting the *in vitro* pregnancy tests. Part of the research was supported by Contract No. 6243-22/6/001 of the Commission of the European Communities.

REFERENCES

Balner, H. (1975). *In* "Breeding Simians for Developmental Biology" (F.T. Perkins and P.N. O'Donoghue, eds), pp. 31-39. Laboratory Animals, London.
Boorman, G.A., Speltie, T.M. and Fitzgerald, G.H. (1974). *J. Med. Primat.*, $\underline{3}$, 269-275.
Goosen, C. (1972a). *In* "Breeding Primates" (W.I.B. Beveridge, ed.) pp. 38-40, S. Karger, Basel.
Goosen, C. (1972b). *In* "Breeding Primates" (W.I.B. Beveridge, ed.) pp. 88-91, S. Karger, Basel.
Hogden, G.D. and Ross, G.T. (1974). *J. Clin. Endocr.*, $\underline{38}$(5), 927-931.
Valerio, D.A., Miller, R.L., Innes, J.R.M., Courtney, K.D., Pallotta, A.J. and Guttmacher, R.M. (1969). *In "Macaca mulatta"* (D.A. Valerio, R.L. Miller and J.R. Maitland-Innes, eds), pp. 70-71. Academic Press, New York.
Valerio, D.A., Leverage, W.E., Bensenhaver, J.C. and Thornett, H.D. (1971). *In* "Medical Primatology 1970" (E.I. Goldsmith and J. Moor-Jankowski, eds), pp. 515-525. S. Karger, Basel.

ADAPTATION OF MONKEYS TO EXTREME CONDITIONS

V.I. CHERNYSHEV

The USSR AMS Institute of Poliomyelitis and Viral Encephalitides, Moscow, USSR.

The problem of adaptation of monkeys acquires paramount significance since monkeys are widely used in medical and biological experiments, particularly for production of antiviral drugs; large contingents of these animals are transported from natural habitats to various parts of the world. It is no secret that in some animal houses of research institutions the mortality of monkeys after their admission at times reaches 50-60%; at the same time in captivity the greatest longevity of monkeys is shown (Jones, 1962) and the highest concentration of animals per unit of floor-space and volume is recorded.

The study of the norm of reaction is one of the methodical approaches to the problem of adaptation. The more complete knowledge we gain about the limits of the norm of reaction in its qualitative and quantitative contents, the more effective will be biotechnical measures directed toward lessening the action of unfavourable factors and compensation of missing factors and the more perfect will be the forecasts of the results of introduction.

To elucidate the limits of adaptational abilities of monkeys, mainly to the climate of northern latitudes (Moscow region), experiments on year-round housing and breeding of monkeys in open-air cages under the conditions of the Moscow region were carried out over many years. From spring 1964 to April 1969 (the last date is accepted as the date of completion of the experiment; in reality the monkeys were kept in open-air cages until October 1971) 7 groups of monkeys, a total of 34 animals, were tested; 24 rhesus macaques and 10 African green monkeys. Thirteen babies were born in open-air cages. The majority of the new-borns reached sexual maturity and produced normal offspring.

We suggest that to develop and study a very complicated biological problem of adaptation it is necessary to consider two sections of the problem independently; adaptation to the climate (*acclimatization*) and adaptation to the weather (*veterization*). When living animals are kept in artificial conditions, adaptation to these conditions is designated as *accommodation*. The term *veterization* allows us to draw attention to complex reactions of the organism which are displayed as a result of extremely changeable and complicated physical conditions of environment united under the term *weather*.

Adaptation of the species to a new climate depends, in the end, on successful veterization.

Tables I and II present data on the duration of housing of monkeys in the open-air cage at different air temperatures and comparative data on adaptational reactions upon veterization, acclimatization and accommodation. In this respect the scheme of acclimatization process of man in the north suggested by G.M. Danishevsky (1956) is of interest. In our opinion this scheme is

TABLE I

Duration of uninterrupted housing of monkeys in the open-air in the Moscow region at different air temperatures

Temperature of the air in the open-air cage	Months	%
above +20°	3.3	5.5
+0.1 +20°	33.7	54.5
0.0 -20°	21.5	35.2
-20.0 -38°	2.4	3.9
	60.9	100.0

useful not only from a medical point of view but from a general biological point of view as well.

With some changes, this scheme can be applied to acclimatization of monkeys in northern latitudes. Instead of three phases, (i) initial phase of acclimatization, (ii) phase of rebuilding of dynamic stereotype, (iii) phase of persistent acclimatization, we introduce two phases for monkeys: (i) initial phase of acclimatization or veterization and (ii) phase of persistent acclimatization on the basis of a new morpho-physiological level. There are two variants in the first phase. 1. Adaptation to the environment of a healthy organism which is brought about by balancing at new functional and morphological levels resulting in the emergence of a new morpho-physiological level. 2. A characteristic of organisms affecting adaptational mechanisms, which is caused by latent and evident disease. This means that the organism must first overcome the disease and then work out necessary adaptations.

It is only thus that one can speak about acclimatization of the organism. Favourable conditions of housing, including a comfortable microclimate, diet and specific treatment, contribute to recovery and serve as an artificial means of preparation of monkeys to acclimatization. Taking into account that the monkeys are heterogeneous with regard to their state of health and their ability to adapt, even healthy monkeys cannot all be used in acclimatization experiments. Monkeys with relatively thin hair, or lacking inborn ability to conserve heat by posture or by mutual warming, are not suitable for acclimatization in extreme conditions. The variety of individuals in the population (according to adaptational possibilities) creates a great variety of responses both to the action of different factors and to a new environment as a whole.

It can be concluded that healthy monkeys are capable of enduring environmental changes over a wide range. In our opinion the above considerations of adaptation of monkeys to extreme conditions can be used, on the one hand, for planning work on acclimatization and, on the other hand, for experimental primate material useful for anthropologists working out the history of distribution of man in various regions of the world.

REFERENCES

Danishevsky, G.M. (1956). Acclimatization of Man (Population). (In Russian.) *In* "Big Medical Encyclopedia", Vol. 1, pp. 422-450. State Publishing House of Medical Literature, Moscow.

Jones, M.L. (1962). *Symp. Zool. Soc. London*, **17**, 427-457.

TABLE II

Adaptational reactions of monkeys to weather (climate) of the Moscow region and to artificial conditions of animal houses

Adaptational reactions	Veterization, Acclimatization	Accommodations (artificial conditions)

Motor activity
  In warm seasons
  In cold seasons
	
	Maximal
	Minimal
		Maintained at the same level year-round

Body weight of adult monkeys	Undergoes seasonal changes	Seasonal changes are not marked
Oxygen consumption	For veterization - reaches maximum in cold seasons. For acclimatization - relatively low in cold seasons; it is associated with heat isolation properties of hair	Seasonal changes are not marked
Heat production	Similar to the reaction of oxygen consumption	Seasonal changes are not marked
Pilomotor function	Is markedly expressed	Is not marked in healthy monkeys
Deposit of fat in subcutaneous cellular tissue	Is markedly expressed	Is not marked in healthy monkeys
Appearance of hair	Downy, brilliant, clean, thick, soft	Felted, colourless, thin, rough, often dirty
Length of hair	Increases upon long-term life in the open-air cage	Remains without changes
Surface of heat emission determined by posture	Minimal in the cold, maximal in the warmth	Remains constant during the year
Skin temperature	In the cold, 5-12° lower	Remains relatively constant
Rectal temperature	In winter, at lower limits of the norm (veterization); in the range of 36-40° (acclimatization)	36-40°

THE ECONOMICS OF NON-HUMAN PRIMATE CONSERVATION

L.B. CUMMINS, G.T. MOORE AND S.S. KALTER

*Division of Microbiology and Infectious Diseases,
Southwest Foundation for Research and Education,
San Antonio, Texas USA.*

Biomedical scientists only recently began serious planning of self-perpetuating non-human primate breeding and research colonies (Bermant and Lindburg, 1973; Goodwin, 1975; Hummer, 1968; Kalter and Moore, 1972; Neurauter and Goodwin, 1972; Schmidt, 1972). Seven breeding systems were defined during this planning cycle (Neurauter and Goodwin, 1972), and scientists today still disagree as to which system is the most efficient and economically feasible.

For the past 17 years, the Southwest Foundation for Research and Education (SFRE) has produced and reared baboons using two of the seven systems: an outdoor colony cage environment and an individual cage situation. Colony performance statistics are available through several sources (Kalter and Moore, 1972; Kraemer et al., 1975; Moore, 1975). These statistics also indicate past chimpanzee breeding colony performance (Kraemer et al., 1975), which has been altered considerably by recent management changes.

This report is presented to demonstrate the actual cost of operation of the SFRE baboon and chimpanzee colonies for the years 1973 through 1976 when available, and to project a trend analysis of the cost of operations through 1978. Costs of infant production using production performance rates and *per diem* cost will also be presented.

Historically, the SFRE has imported a large portion of its animals, either directly from Africa via its own trapping facility, or through commercial suppliers. Since 1974, however, all animals have been purchased through commercial sources.

The base animal cost is the one factor over which the investigator has no control and this cost will eventually become unacceptable enough to force estic production, all moral issues aside. Table I shows actual animal (baboon, delivered in San Antonio) costs for 1973 through June 1976. The 1973 figures are abnormally low since they reflect the purchase of animals of the less than 5 kg weight range only.

Table II shows the constant rise of local quarantine costs. Percentage increases have remained stable, with a slightly decreasing figure each year. These cost figures reflect local management efforts and represent approximately one-fourth of the new animal costs.

Conditioned animal cost for a ready-to-use research subject. Table III shows actual cost of a baboon through the middle of 1976; as in Table I, the 1973 figure was slightly low because of the small-sized animals purchased in that year.

Accurate purchase costs for chimpanzees were available only for 1975.

TABLE I
Base animal cost (baboon)

Year	Actual (dollars)	Projected (dollars)
1973	88	
1974	250	
1975	260	
1976	309*	
1977		395
1978		462

*Through June 1976

TABLE II
Quarantine cost (baboon)

Year	Actual (dollars)	Projected (dollars)
1973	65	
1974	76	
1975	88	
1976	101*	
1977		113
1978		125

*Through June 1976

TABLE III
Conditioned animal cost (baboon)

Year	Actual (dollars)	Projected (dollars)
1973	153	
1974	326	
1975	348	
1976	410*	
1977		528
1978		613

*Through June 1976

Purchase price per animal was $1,300. Quarantine expenses for six months produced a cost per conditioned animal of $1,924.

An intangible expense, yet to be considered, is the price of a disease outbreak in a group of quarantine animals. A specific tuberculosis outbreak in 1975 cost the SFRE $6,304. Not included in this cost were the extra 60 days lost in resolving the problem, the extra time lost replacing the animals involved, and the extra cost and time expended by the research program for which these baboons were destined. The $6,304 divided among the survivors of the outbreak, raised the cost per conditioned animal by $573. A similar outbreak

TABLE IV
Per diem cost (baboon)

Year	Actual (dollars)	Projected (dollars)
1973	0.58	
1974	0.75	
1975	0.94	
1976		1.12
1977		1.30
1978		1.48

TABLE V
Infant produced cost (baboon)

Year	Actual (dollars)	Projected (dollars)
1973	255	
1974	330	
1975	413	
1976		491
1977		570
1978		650

TABLE VI
Chimpanzee per diem

Year	Actual (dollars)	Projected (dollars)
1973	1.49	
1974	2.26	
1975	2.87	
1976		
1977		
1978		

in 1976 has yet to be calculated.

Colony-wide infant production costs are difficult to determine because of the dynamics of the SFRE colony; however, several baboon breeding units remain intact enough to provide accurate data. A typical unit consists of two males and 24 females. As reported by Moore (1975) and by Kraemer and Vera Cruz (1972), adult female baboons, after production of the first infant in captivity, will produce 0.9 infants per adult female year. These performance figures combined with Table IV (baboon *per diem* costs) yield infant produced cost figures as indicated in Table V.

Chimpanzee production performance rates for 1967 through 1972, as published by Kraemer and Vera Cruz (1972), show 0.49 term pregnancies per adult female year. Management technique changes have raised this performance to 0.71 term pregnancies in the 1974-1976 period. Applying this rate to 1975 chimpanzee *per diem* costs (Table VI), yields an infant produced cost of $2,002 in 1975.

In conclusion, we have recorded a steady cost increase in all areas of colony management. Projections indicate a continued increase in the immediate future. Many factors, which are beyond the control of the scientific community, can and will affect these projections. Our goal is to reduce the costs and increase the production rates. The more endangered the non-human primates become, the greater the probability that the colony-reared infant will become the least expensive source of laboratory non-human primates, if not the only source.

ACKNOWLEDGEMENT

This laboratory serves as the NIH/WHO Collaborating Center for Reference and Research in Simian Viruses.

REFERENCES

Bermant, G. and Lindburg, D.S. (1973). *J. Med. Prim.*, 2, 324-340.
Goodwin, W.J. (1975). *Lab. Anim. Handbook*, 6, 151-156.
Hummer, R.L. (1968). Pan American Health Organization, Sci. Pub. No. 182, pp. 126-131.
Kalter, S.S. and Moore, G.T. (1972). *In* "Medical Primatology 1972" (E.I. Goldsmith and J. Moor-Jankowski, eds), Part I, pp. 105-114. S. Karger, Basel.
Kraemer, D.C. and Vera Cruz, N.C. (1972). *In* "Breeding Primates" (W.I.B. Beveridge, ed.), pp. 42-47. S. Karger, Basel.
Kraemer, D.C., Kalter, S.S. and Moore, G.T. (1975). *In* "Breeding Simians for Developmental Biology" (F.T. Perkins and P.N. O'Donoghue, eds), pp. 41-47. Laboratory Animals, London.
Moore, G.T. (1975). *Lab. Animal Science*, 25, 798-801.
Neurauter, L.V. and Goodwin, W.V. (1972). *In* "Breeding Primates" (W.I.B. Beveridge, ed.), pp. 60-75. S. Karger, Basel.
Schmidt, L.H. (1972). *In* "Breeding Primates" (W.I.B. Beveridge, ed.), pp. 1-22. S. Karger, Basel.

THE BREEDING OF NON-HUMAN PRIMATES FOR
BIOMEDICAL RESEARCH

J. BLEBY

*Medical Research Council,
Laboratory Animals Centre, Carshalton.*

BACKGROUND HISTORY

Until 1950, when tissue culture techniques were developed as a basis for virology, laboratory demands for primates were few and transport was in small consignments, usually by sea. However, the use of primates as sources of tissue led to a rapidly increasing demand, particularly when virus vaccines became a commercial possibility.

This rapid expansion has had numerous undesirable results, including trapping animals on a large scale, collecting them together and transporting them by air, sometimes in bad conditions, which have been regulated by commercial rather than scientific considerations. At the same time these demands have coincided with an increasing interest in the use of primates for other fields of biomedical research where their similarity to man is important.

Developments have also occurred at an international level especially in the U.S.A. and U.S.S.R. In Georgia, U.S.S.R., the Sukhumi Institute of Experimental Pathology and Therapy has a primate centre with origins dating back to 1927 and the rapid expansion of experimental primatology over the last 20 years has resulted in the Centre being extended to house several thousand primates of several species.

In the U.S.A., a comprehensive programme of regional primate centres culminated in the Oregon Primate Research Centre being founded in 1960 and six other Regional Centres in 1962. These Centres are financed by the National Institutes of Health.

In Europe, the Dutch National Primate Centre at Rijswijk is so far the only established national centre, although Germany, France and Italy are also considering various schemes. It would appear that the only attempt to co-ordinate private interests in the U.K. at a national level was made during the 1960's by Dr. J.R. Napier, Professor P.L. Krohn and the late Professors K.R. Hall and G.W. Harris. It seems that no proposals were formulated although a plan was considered in 1966 to establish some form of national primate centre at Oxford University.

As the demand for primates grew and their availability from the wild decreased, the Medical Research Council agreed that its Laboratory Animals Centre (LAC) should "make an investigation into the existing provision and future need for laboratory primates". The survey was conducted by the LAC Primate Officer, Mr. K.R. Hobbs between 1969 and 1971, and a report was published in 1976 entitled "Laboratory Non-Human Primates for Biomedical

Research in the United Kingdom - A Report to the Medical Research Council on the Existing Provision and Future Needs" (Hobbs and Bleby, 1976).

MEDICAL RESEARCH COUNCIL SURVEY

Information was collected on the following:

- why primates are needed;
- the supply, quarantining and conditioning of wild trapped primates including their export and import and the associated legislation and procedures;
- the problems of conservation;
- the problems associated with the use of wild trapped primates for research purposes;
- the economics of supplies from the wild and their use;
- the desirability and need for specialised primate research facilities;
- the feasibility of breeding primates for research;
- the role of overseas stations;

The main findings of the report were as follows:

1. There is a continual and growing demand for primates in biomedical research over an ever widening field.
2. Over 90% of primate supplies come from the wild even though some sources are rapidly running down or have become non-existent.
3. In the U.K. 85% of all primate work involves four species only, and 94% involves seven species only (Fig. 1).

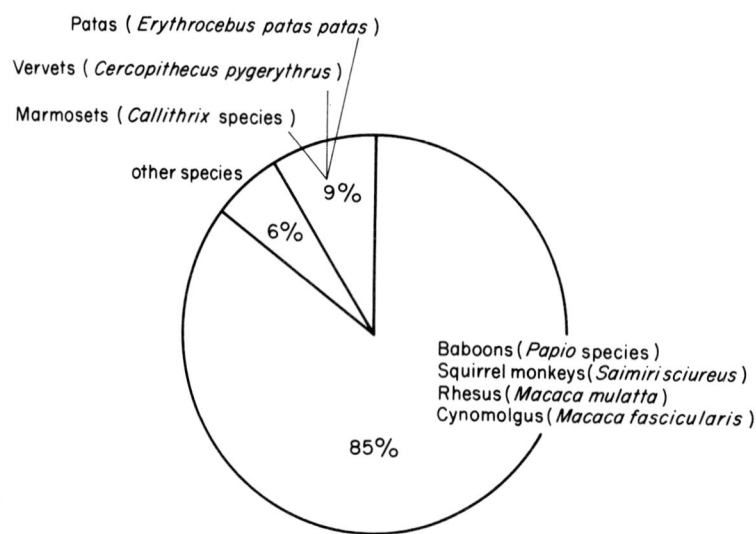

Figure 1.

4. Nearly all wild monkeys either are diseased, or have been, often by infections that are not only communicable to man, but also seriously interfere with experimental work.
5. The solution to many of the problems is to breed primates specifically for biomedical use.

BREEDING AND ITS ADVANTAGES

All the species required by the U.K. are relatively easy to breed and two, the marmoset (*Callithrix jacchus*) and the squirrel monkey (*Saimiri sciureus*) require accommodation comparable with that used for many common laboratory animals. The baboon because of its large size and late maturity may present some practical problems and is therefore, like other large primates, not so suited for laboratory use as the smaller species, although there are occasions when the baboon has to be used, e.g. as a natural host to bilharzia.

Occasionally, requests are made for unusual species, such as anthropoid apes. The chimpanzee is scarce in the wild and is therefore now protected by many countries, import licences for their experimental use being refused.

The World Health Organisation (1971) has stated that "The only satisfactory solution for ensuring the future supply of adequate numbers of suitable monkeys and apes for laboratory use is to breed them in captivity". Throughout the survey, all those interviewed agreed with the principle of breeding for research.

The advantages of breeding primates may be summarised as follows:-
(a) *Quality* The production of disease-free animals increases the validity of research and eliminates the hazards to personnel which are serious, e.g. B virus, Marburg agent, tuberculosis, salmonellosis and shigellosis.
(b) *Standardisation* The supply of uniform groups of animals of known biological parameters results in more accurate work and leads to a reduction in the number of primates needed for a particular study.
(c) *Availability* The provision of a constant source of animals ensures continuity of supply which saves time and money. It also guards against the restrictions brought about by conservation measures.
(d) *Suitability* The use of primates can be extended to include work demanding pregnant, foetal and young animals.
(c) *Ethics* Some moral objections are removed as there is no need to use captured wild animals.

CONCLUSIONS

As the reasons for breeding primates for research would appear to be overwhelming, why has it not already come about? The answer is almost certainly because of the misguided belief that it is cheaper to use a wild animal. However, when it is realised, for example, that the common marmoset (*C. jacchus*) becomes sexually mature at 17 months of age and produces, on average, three young per female per year, the price of a bred animal becomes comparable with the cost of breeding other animals for research.

It is appreciated, that the marmoset is unlikely to meet the requirements of all research workers and a number of larger primates will continue to be required. But it is probable that these requirements will be in the minority, especially if some workers investigate the smaller species instead of continuing to use the larger species like the rhesus, which probably owes its popularity to the fact that it used to be the most prolific in the wild.

There would appear to be no scientific reason why the majority of laboratory primates cannot be bred for the purpose, and some suppliers have already started to establish breeding colonies. It is now up to the users to realize that it is uneconomic to continue using wild animals and to place contracts with their suppliers for purpose bred animals.

Finally, it should be mentioned that government licensing authorities are likely to begin to apply the same minimum standards to laboratory primates as they are already applying to other commonly used species. These standards can only be achieved by breeding, which will also safeguard supplies as an increasing number of wild sources cease to exist.

REFERENCES

Hobbs, K.R. and Bleby, J. (1976). "Laboratory Non-Human Primates for Biomedical Research in the United Kingdom - A Report to the Medical Research Council on the Existing Provision and Future Needs", Medical Research Council Laboratory Animals Centre, Carshalton, Surrey.

World Health Organisation (1971). Health, Aspects of the Supply and Use of Non-Human Primates for Biomedical Purposes. Technical Report Series No. 470, p. 11, W.H.O., Geneva.

RESEARCH MANAGEMENT OF A PRIMATE ANIMAL LABORATORY
FOR MEDICAL EXPERIMENTATION BY CLINICIANS

J. MOOR-JANKOWSKI[*] AND E.I. GOLDSMITH[**]

*Laboratory for Experimental Medicine and Surgery in Primates (LEMSIP)
of New York University Medical Center.
**New York Hospital-Cornell Medical College, New York.

The Laboratory for Experimental Medicine and Surgery in Primates (LEMSIP) was established in 1965 by the joint effort of medical scientists, mainly research oriented clinicians, in need of access to primate animal experimentation. It is a unit of the New York University Medical Center which administers it on behalf of the Associated Medical Schools of New York. The Association represents 12 medical schools in New York State and the New York Academy of Medicine; however LEMSIP serves also investigators nationally and internationally.

The major objective of the laboratory is to assist medical researchers who carry clinical patient responsibilities and who represent more than 60% of its users. Therefore, the research management of the laboratory has been developed towards responsiveness to the needs of physicians who cannot forgo their daily patient responsibilities and yet want to maintain close supervision of their research projects.

FACILITIES AND PROCEDURES

The operation and the design of LEMSIP are closely interdependent. The physical plant, the caging systems and the methods of husbandry and experimentation have been developed without relying on the previously existing concepts (Davis and Moor-Jankowski, 1970). Chimpanzees, baboons and macaques are maintained under conditions derived both from human hospital practices and from husbandry methods of the modern poultry breeding facilities. This has resulted in 12 years of operations with extremely low morbidity and mortality and complete absence of epidemics (Moor-Jankowski et al., 1970).

Since 1965, the personnel health policy of the laboratory requires the wearing of complete protective clothing when in contact with animals, physical examination and TB testing every six months, and additional tests related to the existing programs as directed by the staff physician (Muchmore, 1975). Investigators are only exceptionally allowed into animal housing. Instead, the animals are brought to the investigator in areas specifically designated for this purpose. This has been made possible by the design of LEMSIP cage systems (Davis and Moor-Jankowski, 1970). Most handling of the animals, as well as surgery and necropsies, are performed by LEMSIP personnel, although visiting surgeons are readily accepted, and specimens may also be processed in outside laboratories. The animals undergo routine quarterly clinical and laboratory health examinations. Monkeys are quarantined for 90 days and chimpanzees for 180 days. Quarantine applies also to LEMSIP animals returning from other institutions.

SERVICES AVAILABLE

1. Professional know-how, space, facilities, animals and technical staff under the general supervision of the Director, closely advised by the professional staff, are provided for the use of investigators at LEMSIP as well as in their own institutions. While we believe that our physical plant offers optimal possibilities for primate research, it is the investigator who decides whether he prefers to utilize our facilities or to receive our assistance in his own institution.
2. Obtaining, examining, conditioning, certifying and delivering healthy primate animals for use within the laboratories of extramural investigators.
3. Obtaining and delivering sera, organs and other biological materials requested by extramural investigators.
4. Training of scientists and of technical personnel, and teaching of students in medical primatology.

RESEARCH MANAGEMENT

The general rules of the LEMSIP research management program are as follows:
1. The investigator has to explain in writing his general objectives and provide a detailed experimental protocol. Standard guidelines and forms are provided for this purpose. Past experience has shown that it is advantageous to have the investigator discuss his research with LEMSIP's Chairman of the Utilization Committee or with LEMSIP's Director prior to the final submission of the protocol. At this stage of planning, sequential and/or simultaneous sharing of the available animals is encouraged.

During the past 12 years no protocol submitted to the LEMSIP Utilization Committee has been turned down; however, some investigators withdrew their proposal after initial discussions, having realized its unfeasibility. Most of the protocols were modified following advice of the Utilization Committee in order to select a biologically more appropriate primate species, to avoid the use of endangered animals, to adapt the experimental protocol to the techniques of primate animal experimentation, to safeguard the humane approach and to decrease the cost of the experimentation.

2. After the protocol has been approved by the Utilization Committee the investigator obtains from LEMSIP's Manager of Business Affairs an exact cost estimate which must be approved by him and his institution.* He then meets with a professional member of the staff and the technicians assigned to his program to whom he is able to entrust as much of the responsibility as is agreeable to him. He retains, however, the intellectual responsibility for the experiment and must be available at all times for telephone consultations.
3. Detailed protocols are automatically mailed to the investigators on each procedure, with copies to LEMSIP's professional person in charge of the program.
4. The investigators may, if they so desire, bring to LEMSIP their own

---

*LEMSIP cost accounting (Davis and Fitzgerald, 1972), was used as the basis for developing national US fiscal guidelines for animal facilities (DHEW, NIH, 1974).

personnel and equipment, or carry out experimentation in their own facilities using animals maintained at LEMSIP and/or LEMSIP's personnel. In order to safeguard their intellectual property the investigators are requested not to divulge more information than necessary for the carrying out of their protocols.

Using the above procedures since 1965, LEMSIP has accommodated more than 500 research projects by collaborating and visiting scientists, mainly from New York and the northeast but also nationally and internationally from 10 countries and 65 institutions.

ACKNOWLEDGEMENT

Supported by USPHS, NIH, Contract RR-2181.

REFERENCES

Davis, J. and Fitzgerald, T. (1972). *In* "Medical Primatology, 1972" (E.I. Goldsmith and J. Moor-Jankowski, eds), Part 1, pp. 37-49. S. Karger, Basel.
Davis, J. and Moor-Jankowski, J. (1970). *In* "Medical Primatology, 1970" (E.I. Goldsmith and J. Moor-Jankowski, eds). S. Karger, Basel.
DHEW, NIH and Association of American Medical Colleges - Division of Research Resources (1974). Cost Analysis and Rate Setting Manual for Animal Resource Facilities. NIH, Washington.
Moor-Jankowski, J., Muchmore, E. and Davis, J. (1970). *In* "Infections and Immunosuppression in Subhuman Primates" (H. Balner and W.I.B. Beveridge, eds). Munksgaard, Copenhagen.
Muchmore, E. (1975). *Cancer Research Monograph Series*, $\underline{2}$, 81-99.

BLOOD GROUPS OF APES AND MONKEYS: THEIR USE AND VALUE
IN EXPERIMENTATION AND BREEDING OF PRIMATE ANIMALS

W.W. SOCHA, A.S. WIENER AND J. MOOR-JANKOWSKI

*Laboratory for Experimental Medicine and Surgery in Primates (LEMSIP),
and WHO Collaborating Center for Haematology of Primate Animals,
New York University Medical Center, New York.*

INTRODUCTION

   Considerable information has been accumulated on blood groups of primate animals during the fifteen years of activities of our Primate Blood Group Reference Laboratory at LEMSIP (for review see Moor-Jankowski and Wiener, 1972). The methodology developed by us (Socha et al., 1972) is based on the standard methods used in typing human blood for clinical and forensic purposes. Actually, the same techniques are used in our primate animal work as in our work on human blood for hospital blood banks and for the Office of the Chief Medical Examiner of New York City, and include all the human blood typing routines, namely saline agglutination, antiglobulin, ficinated red cell technique, and inhibition tests on saliva. The results obtained are as sensitive, specific and reproducible as those required for human patients.
   The knowledge of blood groups is a prerequisite for transfusion therapy in sick animals and in surgery, for transplantation experiments, and in breeding to avoid the dangers of maternofoetal incompatibilities, and to provide genetic markers for selective breeding.

TRANSFUSION

   Apes, with the exception of gorillas, have in their serum preformed naturally occurring antibodies of the human-type (Moor-Jankowski et al., 1964) A-B-O blood group system and the reciprocal blood group antigens on their red cells (Moor-Jankowski and Wiener, 1972). Thus, they have to be tested for their A-B-O blood groups and transfused accordingly; otherwise severe transfusion accidents will occur.
   All primate animals have to be cross-matched for the possible presence of spontaneously occurring antibodies (Socha et al., 1975). If a single transfusion is planned, cross-matching and A-B-O testing will suffice. If there is a chance that the transfusion will have to be repeated, however, the animals should also be tested for their simian-type blood groups (Moor-Jankowski and Wiener, 1965) and transfused accordingly to all their blood group types.
   In Old World monkeys, the human-type preformed anti-A and anti-B antibodies are not associated with A and B agglutinogens on their red cells; these antigens, however, are present on their tissues and in secretions, so that incompatible transfusions and organ transplantation should be avoided. Human-type blood groups of all commonly used apes and monkeys are well-known (Moor-Jankowski

and Wiener, 1972) and the simian-type blood groups are well-investigated in chimpanzees (Kratochvil, 1972) gibbons, baboons, rhesus and crab-eating macaques (Moor-Jankowski and Wiener, 1972).

Tests of primate animal bloods for human-type A-B-O blood groups *cannot be carried out with the unabsorbed human blood grouping sera* available in blood banks. Such testing was attempted in many laboratories in emergency situations and led regularly to failure: the testing reagents must be absorbed or appropriately diluted and control simian red cells must be used, as is required in human blood typing.

BREEDING

Erythroblastosis foetalis used to be a significant cause of foetal wastage, neonatal mortality and mental retardation in man. At present it has virtually disappeared in scientifically advanced countries, thanks to the immunoprophylaxis. In primate animals (Wiener et al., 1975) the conclusions to be drawn from the data presently available indicate that erythroblastosis does not play a role in reproduction of rhesus monkeys (Sullivan et al., 1972) and of crab-eating macaques (Wiener et al., 1975). However, more experimental studies on a larger number of macaques are needed to confirm the data presently available. On the other hand, stillbirths, newborn deaths and newborn morbidity were demonstrated in marmosets (Gengozian, 1969) and in baboons (Verbickij, 1972) and erythroblastotic morbidity was observed in chimpanzees (Wiener et al., 1975). Therefore, breeding programs for these species should include blood typing and matching of parents for their human-type and simian-type blood groups. This is being done in the chimpanzee breeding program at LEMSIP.

For breeding situations where more than one male is available to several females, it is important to remove the low-fertility or infertile males. This can be best accomplished by determining the paternities of offspring produced, using blood group tests, as routinely done for disputed paternities in man, as well as for breeding purposes in horses and cattle.

With the advent of large-scale breeding of primate animals for biomedical research, the need will soon arise for *breeding out* of disadvantageous genetic traits, or for *breeding for* certain desired characteristics. A case in point is here the genetically determined hypercholesterolaemia in squirrel monkeys (Clarkson et al., 1971). Breeding for production of hypercholesterolaemic squirrel monkeys, if done by the harem method, will require culling not only of low-fertility males but also of those whose offspring will demonstrate a low Mendelian segregation for the hypercholesterolaemic characteristics. Thus paternity determination in such troops of animals will again be necessary and could be optimally performed by blood grouping. Our WHO Collaborating Center for Haematology of Primate Animals at LEMSIP, New York, is available for all necessary human-type and simian-type tests and we are often able to provide and ship out matched chimpanzee, baboon and macaque blood for transfusion.

ACKNOWLEDGEMENT

Supported by USPHS, NIH, Grant 12074, Contract RR-2184.

REFERENCES

Clarkson, T.B., Lofland, B.H., Bullock, B.D. and Goodman, H.O. (1971). *Archs.*

Path., 92, 37-45.
Gengozian, N. (1969). *Nature, London,* 209, 722-732.
Kratochvil, D.H. (1972). ed. "Chimpanzee: Immunological Specificities of Blood". *Primates in Medicine,* Vol. 6. S. Karger, Basel.
Moor-Jankowski, J. and Wiener, A.S. (1972). *In* "Pathology of Simian Primates" (R.N. T-W-Fiennes, ed.), Part I, pp. 270-317. S. Karger, Basel.
Moor-Jankowski, J. and Wiener, A.S. (1965). *Nature, London,* 205, 369.
Moor-Jankowski, J., Wiener, A.S. and Rogers, C.M. (1954). *Nature, London,* 202, 663.
Socha, W.W., Wiener, A.S. and Moor-Jankowski, J. (1972). *Transpl. Proc.,* 4, 107-111.
Socha, W.W., Wiener, A.S., Moor-Jankowski, J., Scheffrahn, W. and Wolfson, S.K. Jr. (1975). In press.
Sullivan, P., Duggleby, C., Blystad, C. and Stone, W.H. (1972). *Abstract. Fed. Proc.,* 31, 792.
Verbickij, M.W. (1972). *In* "The Use of Nonhuman Primates in Research on Human Reproduction" (E. Dicfalusy and C.C. Standley, eds). WHO Symposium, Sukhumi 1971 (WHO Research and Training Center on Human Reproduction, Karolinska Institute, Stockholm).
Wiener, A.S., Socha, W.W. and Moor-Jankowski, J. (1975). In press.
Wiener, A.S., Socha, W.W., Niemann, W. and Moor-Jankowski, J. (1975). *J. med. primatol.,* 4, 179-187.

BUY OR BREED?

W. LANE-PETTER

formerly *Clinical School Animal House,
University of Cambridge.*

Biomedical research and other laboratory uses represent by far the greatest of all demands for primates, and it has of necessity fallen on the commonest or most easily obtained species, such as the rhesus or the squirrel monkey. Evidence is provided in the foregoing papers that even a species as numerous as the rhesus has been so heavily predated that it could be in danger of extermination, and that continued collection can no longer be permitted, at least on the present scale.

Further evidence indicates that such collection is wasteful, because of incidental losses in trapping and transportation; that it is dangerous, because of infections that wild caught animals may carry or acquire and transmit to man; and that it may in some circumstances be inhumane and degrading. It is not only those remote from the laboratory - the trapper, the porter, the airline handler - that are at risk. There have been enough tragedies among laboratory workers to remind even the most thoughtless that dangers exist and that they are avoidable.

The choice of species has been dictated largely by ready availability, but there is no reason to suppose that this corresponds at all closely with an ideal choice based on scientific requirements; in fact it would be surprising if this were so. If a primate is required in the laboratory - and there seems to be no doubt that for some purposes a primate is indispensible - then the right primate should be chosen and, like the laboratory rat, developed by breeding and selection. This will take a long time, because the generation interval in all primates is much longer than in rats, but a start has already been made with rhesus, baboons, squirrel monkeys, marmosets, tree shrews and even apes. In the long run such breeding must be the main, even the exclusive, source of laboratory primates. Collection from the wild will have to be rigorously restricted to obtaining very small samples to screen for potential usefulness, or as foundation breeding stock.

On the economic side, the current and projected costs of breeding some species are set out in some detail in this volume. The cost, in cash terms, appears to be high, as all laboratory animal breeding appears to be, relatively. But against the cost of breeding must be set, not only the cash cost of collection, transportation and delivery to the laboratory (including screening and conditioning), but also the hazards of unwanted infections, the lack of uniformity and the social maladjustments in groups of animals formed from wild caught batches, the political problems that sometimes arise in the countries of origin, where conservation needs will often conflict with the business of collection, and the restriction of choice of species to those few that are readily available. All these, and many other factors incidental to the trade, when taken together may load the true cost to the point where breeding turns out to be the most economical method of supply.

These are some of the main reasons why special breeding must today be regarded as inescapable. It is encouraging to see that it is already being undertaken in several countries. The next question that arises is where such breeding should ideally be located. Should it be in the native country of the species being bred, where the climate is presumably right and the food sources appropriate, but where long haul transportation remains a hazard and an added expense for the bench-ready animal? Should it be in the place of use, where climate may be less suitable but transportation of little consequence? Or perhaps in some other place, such as a remote island, which offers the best opportunities for health and population control? All these options are already being tried out, but it is too early to pick one as being preferable to the rest.

For the present, any policy of primate procurement must take account of the following considerations.

1. Collection from the wild of large numbers of primates, even of the commonest species, must in the interests of conservation if for no other reason stop now or in the very near future. Biomedical research cannot override all other considerations.

2. Use of primates in the laboratory must be confined to those purposes for which other species have been tried and found to be inadequate. The primate should be the last, not the first, choice of species.

3. Granted that breeding is necessary, the choice of species to be bred must be related to present and future uses. It may be that some uncommon species that are not used today, because they are difficult to obtain from the wild, will prove to be the most useful laboratory species.

4. In comparing the cost of breeding with that of using wild caught animals, full account must be taken of all the contingent expenses of collection, many of which are apt to be omitted from current balance sheets.

TO BREED OR NOT TO BREED

(with apologies to Hamlet, Prince of Denmark)

To breed, or not to breed; that is the question:
Whether 'tis nobler in the lab to suffer
The slings and arrows of outlandish ailments,
Or to plunge headlong in a sea of troubles
And by own breeding end them? To buy, to breed;
No more; and if to breed we say we end
The infections and the thousand maladies
That purchased flesh is heir to, 'tis a state
Devoutly to be wished, seldom achieved.
To free of pathogens; perchance to dream
That there's no bug. What dreams may come

When we have shuffled through post mortems all
And vainly sought the lesions in our stock
That make calamity of cagèd life.
For who would bear the whips and scorns of time,
The professor's wrong, the director's contumely,
The pangs of déspised care, the law's delay,
The insolence of office myrmidons
That patient merit from the worthy takes,
When he himself might his quietus get
From a bare infected bodkin? Who would bear
The fardels of a breeding colony,
But for the dread of intercurrent death
Of undiscovered nature? What disease,
Hitherto new to science, puzzles the will
And makes us rather bear the ills we have
Than buy in others that we know not of?
Thus ignorance makes cowards of us all,
And thus gnotobiotic derivation
Too soon is sicklied o'er with pathogens
And epidemics of great pith and moment.
With this regard, our programmes turn awry,
And lose the name of science.

Further references added in proof to the chapter by J. Trollope, continued from page 250.

Gouzoules, H.T. (1974). *Primates*, 15, 2-3, pp.287-292.
Lang, E.M. (1972). *The Brit. Vet. Journ.*, 128, 9, pp.433-438.
Macdonald, G.J. (1971). *Fert. Ster.*, 22, 6, pp.373-377.
Maple, T., Erwin, J. and Mitchell, G. (1973). *Primates*, 14, 4, pp.427-428.
Napier, J.R. and Napier, P.H. (1967). *Handbook of Living Primates*. Academic Press, New York and London.
Short, D.J. (1968). *Symp. Zool. Soc. Lond.*, 21, 13, pp.13-20.
Trollope, J. and Blurton Jones, N.G. (1972). *Primates*, 13, 2, pp.229-230.
Trollope, J. and Blurton Jones, N.G. (1975). *Primates*, 16, 2, pp.191-205.
Valerio, D.A., Miller, R.L., Innes, J.R.M., Courtney, K.D., Pallotta, A.J. and Guttmacher, R.M. (1969). "Management of a Laboratory Breeding Colony, *Macaca mulatta*." Academic Press, New York and London.
Van Wagenen, G. (1972). *Journ. Med. Primat.*, 1, 1, pp.2-28.
White, R.J., Blaine, G.A. and Blakely, G.A. (1973). *Amer. J. Phys. Anthrop.*, 38, 2, pp.189-194.